# Storytelling Charts: Visualize Vertical Logic in PowerPoint

A Step-by-Step Guide and Software to Speedy Impactful
Presentations

BY

SAM SCHREIM

# Important

## Unlock Every Advantage Before You Begin: Download the Visual Companion Pack

You'll get far more from these pages when the ideas leap off the paper and onto your screen. That's why I've created a Visual Companion Pack loaded with more **than 70+ pages** of crisp, high-resolution versions of every chart, diagram, and table in the book.

But visuals are only the start. To put each technique straight to work, I'm also including a **free PowerPoint add-in** that lets you build persuasive data stories right inside PowerPoint, no extra software required.
**Install it here:**

www.storytellingwithcharts.com

**Ready to enlist an AI co-pilot?** Grab the e-book at

https://stc.how/aics

— it comes with a quick-reference cheat sheet and 100-plus pages of plug-and-play prompts for the LLM of your choice, accelerating every storytelling chart you create.

With these dynamic digital assets at your fingertips, you'll follow along in real time, watch dull datasets transform into memorable narratives, and see intricate insights become impossible to forget.

**Download your complete visual learner's package now:**

https://stc.how/comp

Let's bring the book into full view—together!

# About the Author

**Sam Schreim** is a global strategy consultant and data storytelling expert who has helped Fortune 500 companies, startups, and government agencies turn complex information into clear, compelling narratives. Over a 20-year career spanning management consulting, venture capital, and executive coaching, he has developed frameworks and tools that bridge logic, design, and persuasion.

He is the creator of **Storytelling Charts Add-In (STCAI)**, a free PowerPoint plugin that democratizes the charting and visual storytelling tools once reserved for elite consultants. Through this book, his workshops, and his software, Sam is on a mission to make persuasive data storytelling accessible to everyone—from students and civil servants to executives and entrepreneurs.

Learn more or access the free visual toolkit at www.storytellingwithcharts.com.

# Table of Contents

# Introduction

*A poet would be overcome by sleep and hunger before (being able to) describe with words what a painter is able to (depict) in an instant. —Leonardo da Vinci*

*The most powerful person in the world is the storyteller who set the vision, values, and agenda of an entire generation that's to come. —Steve Jobs*

Storytelling is arguably the most powerful tool a human being has. As highly social creatures, we use stories to connect with other human beings on emotional and intellectual levels. In doing so, we create bonds between ourselves and others. Through storytelling, we create relationships, friendships, families, and even entire communities. Politicians tell stories to their voters about their visions. Brands tell stories to their consumers about who they are. As individuals, we even tell ourselves stories about who we are to motivate, teach, and entertain ourselves. According to Yuval Noah Harrari, it was storytelling that gave us human beings the ability to create belief systems, relate history, and even dream of the future (Harari et al., 2018). All this is because there's power in a good story. Think back to your childhood. Odds are there are several stories that you grew up hearing and still remember to this day.

There are, in fact, multiple kinds of storytelling, and visual storytelling is one of them. Dating back to the stone age with their cave paintings, visual storytelling is probably the most impactful form of storytelling. This is because visuals convey a richer experience than text-heavy methods of communication. It's also because they communicate information a lot quicker than paragraphs of writing do. This is one of the main reasons why storytelling with charts—that is to say, storytelling using impactful charts–is an incredibly effective way of communicating information. This is especially true in a day and age where the average person visits 89 websites per day and has seven different social media accounts in a quest for instant information (Roothman, 2018).

This book focuses specifically on the vertical logic aspect of storytelling with charts (STC) – the art of crafting individual, persuasive slides that effectively communicate your data story. STC is a very powerful and strategic method of communication that's been around for nearly 60 years. Despite this, few people have been able to truly master it. Having given presentations throughout your academic career and work life, you might think that you are one of those few. The unfortunate truth, however, is that you most likely are not. But you can learn to attain such mastery with dedication and time. Attaining this mastery requires understanding which visuals to use and how to use them according to the STC method. Our surveys and studies show that using powerful visuals that meet all STC requirements, like in the example below, increases audience engagement by 80%. It also increases audience retention by 65% and story believability by a whopping 99%!

As an example, consider the two flags in the image below. These flags would look identical when they're hanging on a pole. There are essentially only two differences between them: one of these products is both more expensive than the other and comes with a story.

| amazon | ebay |
|---|---|
| American Flag 3x5Ft, Embroidered Stars 3'x5' USA Flag Outdoor Heavy Duty and Double Edge Sewing | Vintage 1940's Dettras Stantest Bunting 48 Star American Flag 3' x 5' |
| Price $5.99 | Buy It Now US $300.00 |
| • Quality Material: American flag constructed with strong material to withstand any outdoor weather<br>• Embroidered Stars: The stars are embroidered using densely filled rich white thread.<br>• The stripes are sewn together with two rows of double stitches for added strength | • This is a 48 star, cotton, US flag made with printed stars, and hand sewn stripes. It was made by the Dettra Flag Co. of Oaks, PA. Stantest bunting was introduced in the 1920's and was used until the 1950's.<br>• Bunting was discontinued as a result of the addition of the 49th star in 1959. This flag has been dated 1942-1945 because of the presence of distinctive wartime grommets |

This is just one example illustrating how a 50x difference between price points of like-for-like products can be attributed to the story-price premium. Audiences tend to be more engaged in stories than facts, and stories are incredibly effective ways of getting them invested in products and ideas. This is why things like content marketing have become major fields of their own. The fact is, stories are powerful because they pack various experiences and beliefs into neat packages and communicate them to both the storyteller and the audience. During this process, the audience's brainwaves actually synchronize with the teller's (Renken, 2020). This phenomenon is known as neural coupling, and it makes it very easy for the teller and audience to communicate shared goals and then actively move toward them. Stories allow for trust to be formed between storytellers and audience members. These individuals can then start building a productive relationship based on that trust (Patterson et al., 2012).

Another interesting effect of listening to stories is that getting to the climactic, stressful, or dramatic moments of a narrative causes your brain to release cortisol. Cortisol is the hormone that's responsible for your fight-or-flight response. It also plays a part in solidifying memories of emotional experiences and thus having them stored in your brain. Meanwhile, getting to the resolution of a story—the part in the narrative where conflicts are resolved, problems are solved, and things begin to wind down—causes the brain to release oxytocin. Oxytocin is a hormone that facilitates social bonding, as well as feelings of contentment, calm, and even a sense of security. When you use STC effectively, then, you allow your audience to experience all of this (Peterson, 2017). Your audience forms a more emotional memory of the story they're listening to and thus become more likely to remember it (Begg et al., 1985). They form a greater bond with it and associate the narrative that's being relayed to them in the STC with feelings of contentment, calm, and security. Considering all this, it's no wonder that Yuval Noah Harrari has claimed that human civilizations and societies could not have been built if we didn't know how to tell stories (Harari et al., 2018).

Although the STC framework applies to both vertical logic (how individual slides are structured) and horizontal logic (ow the overall presentation flows), this book is deliberately focused on mastering vertical logic — that is the foundation for telling compelling data stories. You'll learn the key building blocks for data storytelling by first learning how to build impactful individual charts and slides. For a deep dive into horizontal logic and the full STC framework, see the companion volume "Storytelling with Charts: The Full Story".

The various effects of storytelling are essentially why customers are willing to pay more for the flag that comes with a story. It's also why creating and telling good, effective stories are crucial in the business world. So, how do you go about creating good stories in the business world? A good story within this context needs to accomplish a variety of things. First, it must effectively and accurately communicate the data and information that need to be communicated. Otherwise, why are you even telling this story in the first place? Second, it must provide its audience with some context, meaning that it must relate the given facts to specific kinds of people (i.e. the target audience), places, and events. It must allow the listeners to derive meaning from that which they're listening to.

## The Anatomy of STC-Based Presentation

Whatever story you're telling in a presentation, your overall goal is to ask something of your audience. The Rule Of Thumb (ROT) of doing so is to use 20% of the message you're giving to ask your audience to take action. The remaining 80% of your message should focus on content that will automatically lead them to take the action you want them to take. Alternatively, it can focus on getting them to look forward to hearing what you will ask them to do when you're done with your presentation. This structure can be compared to that of an orchestra which builds up the crescendo of a movement by adding one instrument at a time. This is exactly what you want to do until the messages you're giving pile on top of one another, making up 80% of your content.

You should only ask your audience for something once you have completed the story and thus reached its climax (Boyd et al., 2020). Your goal here is to leave them wanting more and then to deliver your pitch.

This is the correct point to deliver your pitch because by now your story will have allowed you to build trust with your audience (Patterson et al., 2012). It will also have helped you to build anticipation, so much so that you no longer have to convince them of your pitch.

In the following chapters, you'll learn what constitutes a message and a story. But before we get into those details, let's define the following terms as they are the two main dimensions that make up the anatomy of an STC-based presentation.

ROT: Think of the story as a series of individual messages that add up to a full story. If we go back to the language analogy, we can think of the messages as the vertical logic. You can think of the words, sentences, and expressions as the horizontal logic that puts them in a sequence that you then use to express yourself.

## Horizontal Logic

A story is typically made up of what are commonly referred to as horizontal logic and vertical logic. Horizontal logic is alternatively called the horizontal flow or flow of the story. How a narrative flows is your starting point when structuring a story, or at least it should be.

Horizontal logic begins with the **goal,** and that goal is used to determine your **strategic objectives**. Your strategic objectives can be further split into main messages, which can in turn be broken down into individual headlines. These **headlines** can then be supported in **charts**, which are then put into a full story deck.

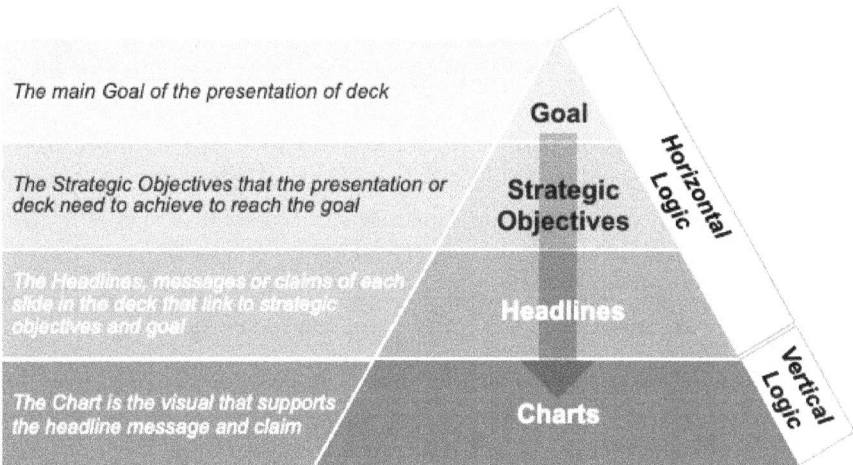

Horizontal logic is the fundamental structure of your storyline and how you can improve it with storytelling hacks, which will, again, be discussed later. Your story deck, meanwhile, is how you substantially enhance your story, using storytelling hacks.

Horizontal logic has several moving parts and has both artistic and creative aspects to it. In contrast, vertical logic is more straightforward, in that it follows a system or a process to reach a conclusion.

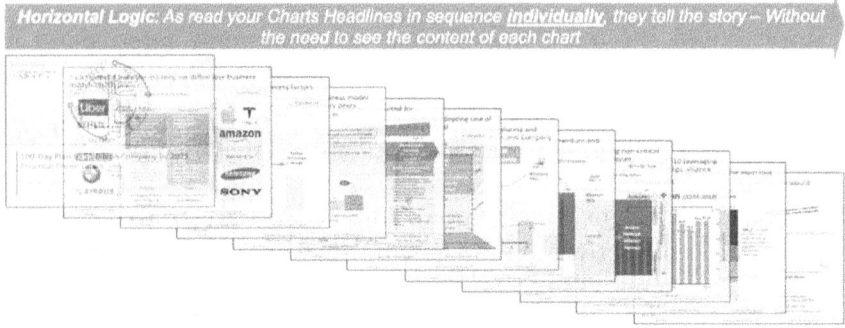

## Vertical Logic

Vertical Logic refers to the individual slides that make up a presentation, often containing charts. These are also known as vertical logic. Vertical logic can often be mistakenly seen as an art, but in reality, there's a great deal more science to it than that. What's more, it can be more easily mastered through a structured approach and system that, when followed, can be very effective. That doesn't mean, however, that you can't be creative when working on vertical logic. You can, as you will see later, in detail. This means that you do not have to sacrifice creativity for science and can retain both within the confines of vertical logic.

An individual chart is known simply as a "visual" element. Such elements are designed to back up given messages, which are usually written at the very top of charts. These header messages have a variety of names, such as the "lead in," "action title" and "caption." As you'll see later, vertical logic charts consist of qualitative charts containing frameworks, conceptual charts, and quantitative charts, which can be enhanced using visual cues.

One thing to remember about charts is that you should follow the One Message Per Chart Rule. This is the one rule to rule all rules about charts.

Put simply, this means that a single chart shouldn't support more than one key insight made in a presentation.

Another thing to remember is that presenting two things on two different slides takes the exact same amount of time that presenting two things on one single slide takes anyways.

## Who Is This Book For?

Everyone, regardless of what profession or field of work they're in, will need to use and/or write presentations at some point. This includes industry professionals, entrepreneurs, consultants, civil/public servants, academics, students, and more. It includes my wife, for instance, who owns her own business and is preparing a pitch for acquiring VC funding and getting a business loan to launch her startup. It includes my 15-year-old daughter who's campaigning to be the president of the school council and is therefore preparing a presentation sharing her campaign promises with the student body and explaining why they should elect her. It even includes my nephew who has had to prepare one presentation after another for his various college courses and is currently preparing one of his thesis defense presentations.

If that's the case, then who is this book for? The short answer? You. The longer answer? This book is for anyone and everyone who needs, uses, writes, and designs presentations or decks for their line of work. It doesn't matter whether you're using PowerPoint, Tableau, Prezi, Keynote, or any other kind of presentation software. If you are someone that's seeking a way to make a case or deliver information to your audience effectively, then this book is for you.

While the STC framework encompasses both vertical and horizontal logic, this book primarily focuses on the vertical logic aspect. Vertical logic deals with the story and logic within individual slides, ensuring that each chart, graph, and visualization effectively communicates your message and supports your overall narrative. On the other hand, horizontal logic is concerned with the overarching structure and flow of your data story, combining individual slides into a cohesive and persuasive narrative. Although both aspects are essential for successful data storytelling, vertical logic is sometimes underemphasized. This book aims to bridge that gap by providing a comprehensive guide to mastering vertical logic and creating compelling, persuasive stories. If you're interested in learning more about horizontal logic and how it complements vertical logic, I recommend exploring "Storytelling with Charts: The Full Story," which covers both aspects in detail.

# The Importance of Vertical Logic

Vertical logic is a crucial component of effective data storytelling, yet it is sometimes overshadowed by discussions of horizontal logic and the overall narrative flow. However, without strong vertical logic, even the most well-structured story will fail to convince and persuade your audience. Vertical logic is what gives your story its credibility, its persuasive power, and its ability to drive action and decision-making. It is the foundation upon which your entire data story rests. That's why I believe vertical logic deserves a dedicated focus and its own book.

By diving deep into the principles and techniques of vertical logic, we can empower storytellers and presenters to create more compelling, persuasive, and action-inspiring stories. This book aims to do just that - to give vertical logic the attention it deserves and to provide readers with a comprehensive guide to mastering this essential skill. Whether you're a seasoned data storyteller or just starting out, understanding and applying vertical logic will take your data stories to the next level and help you achieve your goals more effectively.

### *Acronyms*

As you will undoubtedly notice, I will be using a variety of acronyms in this book that are specific to this framework. These acronyms are intended to help you retain important concepts, hacks, tips, and tricks once you have finished reading *Storytelling with Charts*. To keep you from getting lost in these acronyms though, here is a partial look-up table of the not so common acronyms. I hope you won't have to refer to it too often, but you always can if you need to.

| Acronym | Spell-Out |
|---------|-----------|
| **ROT** | Rule of Thumb |
| **STC** | Storytelling With Charts |
| **HL** | Horizontal Logic |
| **VL** | Vertical Logic |
| **5SUF** | 5-Step Universal Framework |
| **TVMA** | Time, Number of Variables, Message Attribute |

# Chapter 1: Getting Started with Storytelling Charts

Choosing the right software is a key first step to efficiently applying the frameworks, principles, and techniques of Storytelling Charts. While this book is applicable to all platforms, having the right software tools on hand can considerably accelerate learning.

Today, numerous options exist to create charts and presentations, which can be broadly divided into four major categories:

- **Data Visualization Platforms,** which are best suited when you have a more complex dataset, often from a data warehouse. Such platforms include tools like Tableau, Power BI, and Qlik, which offer advanced data visualization capabilities. These platforms excel at data exploration and analysis, but they are overkill for the average business presentation. Their robust capabilities require longer learning curves and, although powerful, can yield over-complex visualizations that are ineffective for storytelling.
- **Premium PowerPoint Add-ins** such as Think-Cell and Empower Suite are examples of tools that enhance the native functionality of PowerPoint with additional charting and time-saving features. These add-ins add professional-grade functionality, but their exorbitant price tags justify the incremental value they provide over modern versions of PowerPoint, especially for storytelling.
- **Light Presentation Apps** such as Apple Keynote and Google Slides can create layouts that have some basic charting capabilities. However, while these apps are excellent for basic presentations, they fall short in providing the crucial features required for advanced data storytelling and vertical logic.
- **AI-Generated Presentation Tools** such as Gamma AI and Beautiful AI (which were popular at the time of writing of this book) leverage artificial intelligence to automate slide design. These tools enable users to create visually appealing presentations effortlessly. However, their reliance on automation often compromises analytical depth. The ease of use comes at the expense of precision and the detailed customization essential for crafting effective vertical logic within compelling data narratives.

All things considered; PowerPoint continues to dominate business presentations for a good reason. The upgraded Office 365 natively includes powerful charting features which have greatly improved in recent years, including many features that were previously only possible through specialized add-ins. The only advantage such premium and paid add-ins offer is enhanced annotation capabilities, but this limited benefit rarely justifies their considerable cost. This is precisely where the Storytelling Charts Add-In (STCAI) becomes relevant.

STCAI focuses on vertical logic, data storytelling, and the most effective visual presentation. This makes it much easier to create charts in accordance with the principles in this book. The add-in is a free download and is compatible with modern versions of PowerPoint. It offers a specialized narrative frameworks and chart visual designs geared towards practice of informative charting. Let's first download and install this tool, which will serve as your practical partner your journey toward effective data storytelling.

In this chapter, I have intentionally refrained from including screenshots of the website or the tool interface. I made this decision because we are continuously upgrading and redesigning both the website and STCAI to improve functionality and user experience. While the visual elements may change over time, the fundamental flow and process described in this chapter should remain consistent. This ensures that the instructions provided remain relevant regardless of the changes and cosmetic updates to the interface.

# Step 1: Download and Installation

## Downloading the Add-In

The first step is downloading the STCAI add-in. To do so, follow these simple steps:

1.  **Visit the Official Website:** Open your preferred web browser and navigate to: **www.storytellingwithcharts.com** Locate the "Download" button prominently featured on the homepage.
2.  **Find your email:** Check your spam folder if you did not receive an email within a few minutes of signing up. Click on the provided link for your operating system, and download the correct version of the add-in.

3. **Click to Download:** Click the "Download" button and wait for the installer file to appear in your downloads folder.

## Running the Installer

With the installer file in hand, proceed as follows:

1. **Open the Installer:** Locate the downloaded installer file. Double-click to launch the installer.
2. **Installation Process:** The installer will guide you through the setup process. Make sure to grant or allow the necessary permissions as this is crucial to enable the add-in to install correctly.
3. **Completion and Confirmation:** This process usually takes just a few minutes. Once installation is complete, you will see a confirmation message. Click "Finish" to exit the installer.
4. **Launch PowerPoint:** Open Microsoft PowerPoint, and a new license window will pop-up. Locate your license, which was sent via email. If you lose or misplace your license that was sent to you in the welcome email, you can always request a new email on the site, and it will be resent to you. Enter the license numbers, accept the terms and conditions, and click verify. You should now receive a confirmation that the license is active.

You should now notice a new tab labeled "STCAI" on the ribbon—this confirms the add-in is successfully installed.

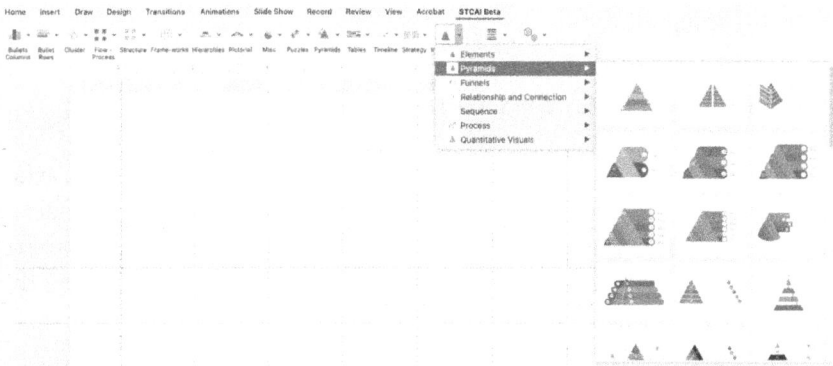

# Step 2: Navigating the Add-In Interface

Now that you've installed the add-in, let explore its simple user

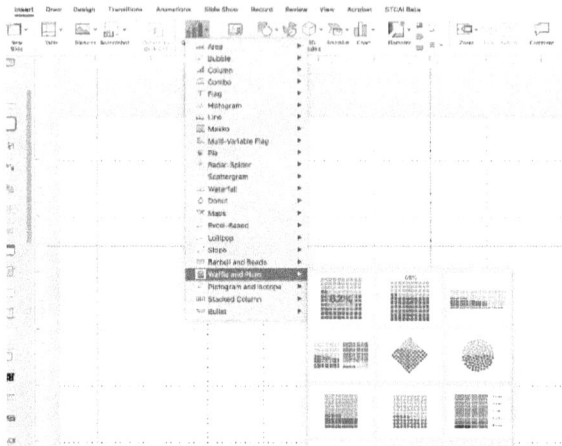

interface. Knowing the layout well will enable you to take advantage of its features fully and speed up your chart creation process significantly.

Please keep in mind that while the interface images shown here may evolve and appear different after add-in updates and the publication of this book, the core functionality and organization described here are likely to remain the same.

The Storytelling Charts Add-In has two interfaces or points of access, seamlessly integrated into PowerPoint.

## *The Quantitative Charts Icon Interface*

Under the "Insert" tab is a new icon named **"Quantitative Charts"** with a lengthy dropdown menu. This menu houses all the formatted quantitative charts available in the codebase, organized by type for intuitive access and usability. Each chart has been developed using the principles in this book.

The charts are distinctive because they natively utilize PowerPoint and Excel functionality, automating much of the formatting process. Every chart implements best practices for data visualization, enabling you to:
- Select the perfect chart for your message with a single click
- Access professionally formatted templates instantly

- Replace sample data in the linked Excel sheet with your own
- Maintain consistent styling across your entire presentation

As of the time of publication of this book, there were hundreds of chart options available within the Quantitative Charts library, with new designs and variations continually being added. From flag charts to line charts, bar charts to plum plots, all the chart types discussed in later chapters have multiple variants to meet particular storytelling requirements.

## The STCAI Tab Interface

The second interface aspect is a dedicated PowerPoint "STCAI" tab in the ribbon. This tab opens a full dashboard, organized into three functional areas:

1. **Qualitative Charts Library:** This extensive collection features 2,000+ qualitative chart templates sorted by categories. From frameworks to process flows, pre-designed elements help you bring concepts to life without hours spent in PowerPoint's Shape Tools. Just click on the relevant category, select the type of chart you would like and add the chart with one click.
2. **PowerPoint Enhancement Tools**: This includes a set of specialized functions that expand on PowerPoint's built-in functionality. Tools for precision alignment, object matching, and text box manipulation can mitigate many limitations in PowerPoint's default toolset. This gives you the ability to fine-tune complex slides without losing the professional look.
3. **AI Assistant**: In addition to the above, there are integrated AI capabilities that utilize sophisticated language models that speed up your vertical and horizontal logic within your content. Although the AI features have some usage limits (it's free with limits, despite underlying costs), it will help you choose the right chart, refine content, and develop your presentation outline.

### Understanding AI Capabilities and Considerations

Functions of the AI assistant include:
- Text enhancement and expansion for headlines and key messages
- Chart type recommendations based on your data and message
- Slide structure suggestions following vertical logic principles

- Data interpretation assistance and insight generation

Although there are a few caveats to watch for (including usage limits on the free version and the limitations of the PowerPoint platform -- in particular, in Office 365 for MacOS), the AI capabilities significantly accelerate the process of learning to create professional-quality presentations.

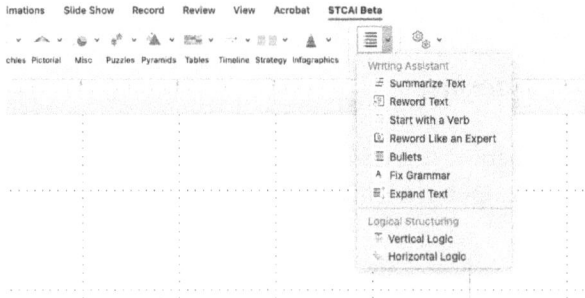

For the most current tutorials and feature demonstrations, visit the resources section of storytellingwithcharts.com, where we regularly update guidance as capabilities evolve. But the most effective approach is hands-on experimentation — many users find powerful workflows just by exploring the available options.

# Step 3: One Click Chart Creation

Now that you are familiar with the interface, let's get create your first vertical slide using the Storytelling Charts add-in.

### Creating a New Slide

Before getting started, make sure you have:
- Prepared your data
- Designed your headline, title, or hypothesis
- Decided what chart type best conveys your message (either through your own analysis, or by using the AI assistant's guidance)

Let's walk you through how to create a slide with a Plum chart:

1. **Choose Your Chart Type**: Go to the Insert tab, click on the "Quantitative Charts" icon, and find the "Waffle & Plum Charts" section.
2. **Add the Chart:** Select the desired Plum chart option. A predefined chart will be inserted in the current slide, filled with sample data formatted in a professional way.
3. **Edit Your Data**: Double-click on the chart to open the linked Excel spreadsheet so that you can replace the sample data with your own values. The data remains linked, ensuring and the chart retains its professional formatting.

### *Refining Your Slide*

When you have your chart and data in place, — it's time to finish the slide by following the principles of vertical logic:

- Ensure your headline clearly communicates the key insight
- Add appropriate annotations to guide the viewer's attention
- Apply any necessary visual cues to emphasize key points
- Verify that the slide follows the best practices discussed in Chapter 3

A key advantage of the Storytelling Charts Add-In is that it handles of the technical aspects of chart generation, allowing you to focus on the strategic parts of your data story. This allows you to focus on crafting the story and ensuring it connects with your audience, rather than wasting time with adjusting colors, alignment, and formatting. In addition to saving hours, as you become more familiar with the add-in, it will enable you to complete more iterations in a single day and ultimately produce higher quality presentations.

## *Recap*

- Consider trade-offs between visualization platforms, premium add-ins, presentation apps, and AI tools when selecting software
- Data visualization platforms (Tableau, Power BI) offer powerful analysis but often create overly complex visuals
- Premium PowerPoint add-ins rarely justify their cost compared to modern PowerPoint capabilities
- Light presentation apps lack robust features for advanced data storytelling
- AI-generated presentation tools create appealing slides but sacrifice analytical precision
- PowerPoint with Office 365 provides robust native charting capabilities

- STCAI enhances PowerPoint through two interfaces: Quantitative Charts icon and STCAI tab
- Quantitative Charts icon provides pre-formatted chart types organized by category
- STCAI tab includes qualitative charts library, PowerPoint enhancement tools, and AI assistance
- AI capabilities include text enhancement and chart recommendations, with some usage limitations
- STCAI simplifies chart creation: select type, insert, replace sample data
- Focus on crafting narrative rather than adjusting design elements

# Chapter 2: Vertical Logic and Storytelling Charts—A Formula for Life

*Logic is the anatomy of thought. —John Locke*

Warning! The following is a spoiler for the movie Benjamin Button:

The year is 1918 when Thomas Button abandons his infant son on the porch of a nursing home. This baby has been abandoned because he was born looking exactly like an elderly man. Luckily, one caretaker at the facility takes pity on him and takes him in. She raises him as her own and duly names him Benjamin. As Benjamin grows in size, he continues to look like the patients populating the nursing home, despite being only a child. Despite this, his physical condition seems to reverse as he grows up. In other words, he ages backwards, getting younger and younger with each passing year. The older Benjamin gets, the younger he looks.

Around the time that Benjamin starts looking at a suitably young age, he meets the granddaughter of one of the nursing home residents and falls in love. Luckily, Daisy, the woman in question, returns his feelings, and the two get married. After Daisy gives birth to their daughter, however, Benjamin leaves them, thinking that he cannot be a suitable parent, given the fact that he's still aging backward. Ten years later, Benjamin goes back to his wife looking like a much younger man, and he and Daisy get back together. At the end of the movie, Benjamin reverts back to a toddler, showing early signs of dementia, before he vanishes from existence completely (Fincher, 2008).

This story, as odd as it sounds, is quite similar to the Lindy effect, which is another kind of reverse aging phenomenon, though there are some key differences.

# Lindy vs. Benjamin Button

Lindy's Law, otherwise known as the Lindy Effect, was first explored in an article published in the June 13th issue of the New Republic. The article, written by Albert Goldman, explained that "the life expectancy of a TV comedian is proportional to the total amount of his exposure on the medium" (Goldman, 1964). The article had gotten its title from a deli in New York City called Lindy's Eatery, where comedians would "foregather every night to conduct post-mortems on recent show biz action."

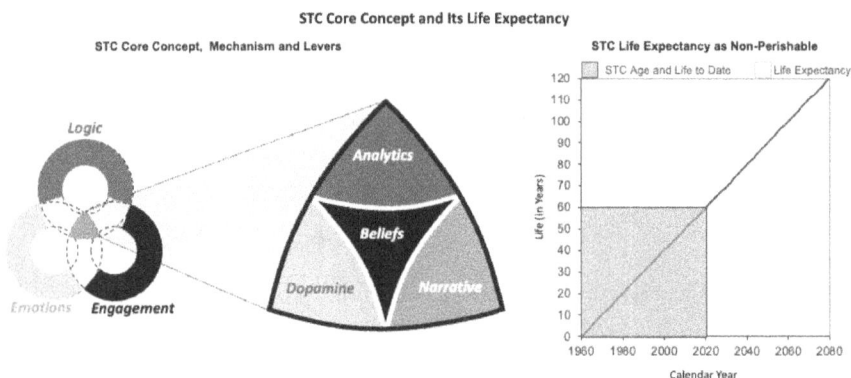

STC Core Concept and Its Life Expectancy

The Lindy Principle, which Goldman laid out in his article, later inspired numerous papers and textbook references and turned into a powerful ROT. In his 1982 book *The Fractal Geometry of Nature*, Benoit Mandelbrot dubbed the phenomenon the Lindy Effect. Mandelbrot claimed that according to the Lindy Effect, the more appearances a comedian made, the more appearances they were expected to create (Mandelbrot, 2006). This idea continued spreading, until it was featured in popularized bestselling books like *The Black Swan* and *Antifragile* by Nassim Nicolas Taleb, who in fact dedicates an entire chapter to it in the latter of the two works.

So, what precisely is the Lindy Effect? Taleb explains the concept as follows (Taleb, 2012):

"The first thing to note about the Lindy Effect is to separate between the perishable and nonperishable. The perishable elements have unavoidable organic expiration dates, such as humans, light bulbs, canned food, etc. Nonperishable would be those with no organic, inevitable expiration date."

To clarify that explanation, basically, the Lindy Effect dictates that certain things really do age in reverse. There are two different kinds of things that do this: perishable and nonperishable things. You can think of these perishable and nonperishable things as objects and content. For example, vinyl records, CDs, or DVDs are objects or technologies that are perishable. Content like music and video, on the other hand, are not. Neither are books nor the bible. A printed copy of the bible might be perishable, subject to the wounds of time as it is. Its content, however, is not and can be expressed in tutorial format or printed in even more copies.

The Lindy Effect is relevant to STC because the framework used in STC is nonperishable and, in fact, shows signs of aging in reverse. This is due to the fact that it has survived, grown, and been in use for decades already and will likely continue to be in use for another 60 years.

What makes the STC framework subject to the Lindy Effect like this is that it is grounded on influencing beliefs (Boudry & Braeckman, 2012). It leverages logic, emotions, and engagement to build stories using analytics, heuristics, and a compelling narrative (Lerner et al., 2015).

## *How Did It Go in 40 BLC?*

Before exploring the STC framework in greater detail, let's recap how building STC-based decks worked in the year 40 BLC (Before Laptops and Computers). Back then, STC followed both horizontal logic, as presentations were made up of individual slides, and vertical logic, as presenters had to create a deck of slides.

The process of creating slides was entirely manual, involving sketching charts on paper, using slide rules for calculations, and having drafters transform the sketches into paper-based slides. Typists would then add text to the slides, and the final product would be shared with clients or produced as overhead transparencies for presentations.

Fast forward to today, and the process has changed dramatically. Computers and laptops have replaced manual methods, and most tools used for designing presentations now have template features and plugin apps that simplify chart creation.

However, despite the advancements in technology and the facelift presentations have received with various colors and icons, the core concept of STC has remained essentially the same over the past 60 years. The process of creating slides has become more efficient, contributing to the growth of the trillion-dollar global management consulting industry by making the process more scalable and viable.

The main takeaway is that the STC concept has been around for a while and isn't going anywhere anytime soon. While the software and format used in STC may change, the concept of telling a story using vertical and horizontal logic will remain unchanged for years to come.

The methods outlined in this chapter worked back then, still work now, and are likely to work in the future. These are lasting skills, and this book is the only one on the market today that condenses these methods into a single, easy read, with a focus on the relevance of vertical logic in the STC framework.

## *Top-Down and Bottom–Up*

The problem I have with the majority of my trainees is that they insist on beginning the process with the data, then moving on to figuring out what the best way to visualize and insert it into the story is. It's true that doing this might feel like it's more intuitive. It's also true that most books, training, and methodologies focus primarily on the VL and brush over the HL and the overall story. But in doing so, they forget that HL is a prerequisite for VL. They also forget that identifying the data needs to be the second to last step in the STC process.

If that's the case, how come this book started with a chapter on VL before moving on for HL? The answer is simple. It's because VL is a prerequisite skill to HL. Hence, when applying the STC methodology, you start with HL first. Afterward, VL determines your chart's blueprint before you go on to mine for data.

As you'll discover later when discussing horizontal logic, this process leads to vertical logic, while horizontal logic defines the kind of analysis and data you need to back up a claim or a message. In keeping with that, the last step in horizontal logic is defining the graphs and charts that are required for the story. This, right here, is the main focus of this chapter. Tackling this last step first, before covering the previous steps, might seem nonsensical at first. But it needs to be done so that you learn how to construct the vertical before moving on to structuring the horizontal.

Comparing that to speaking a language, you use the words to construct sentences. The fact is, a sentence cannot exist without words, just as words cannot be made sense of if they're not put into context in a sentence. Considered within those definitions, the HL of your presentation can be thought of as "the sentence." The VL, on the other hand, can be thought of as the words. You don't need to start with the data, that is to say the letters, when you're learning the words, nor do you need to let the letters guide you to what you want to say.

What all this means is that you need to move on to working on your charts only after you've identified everything you need. Developing a slide is a process that unfolds on two levels. Level 1 is what's derived from horizontal logic. Level 2 is the blueprint. Both these levels are conducted from the top-down, not the bottom-up. That means that you need to approach these levels by figuring out your plot first and collecting your data after. That way you will be able to hone in on what's relevant to you without having to go on a data mining sprint.

## The Anatomy of STC

How exactly should you go about building your story with charts?

Rule number one is that you should always avoid working backward in STC.

The right strategy to adopt when building a story with charts is to begin by figuring out what your goal, strategic objectives, and big idea are. These will help you to identify the issues that set the framework of your storyline. They will make it possible for you to establish which claims and headline messages should be stated to articulate best-guess hypotheses.

# The STC Framework Method

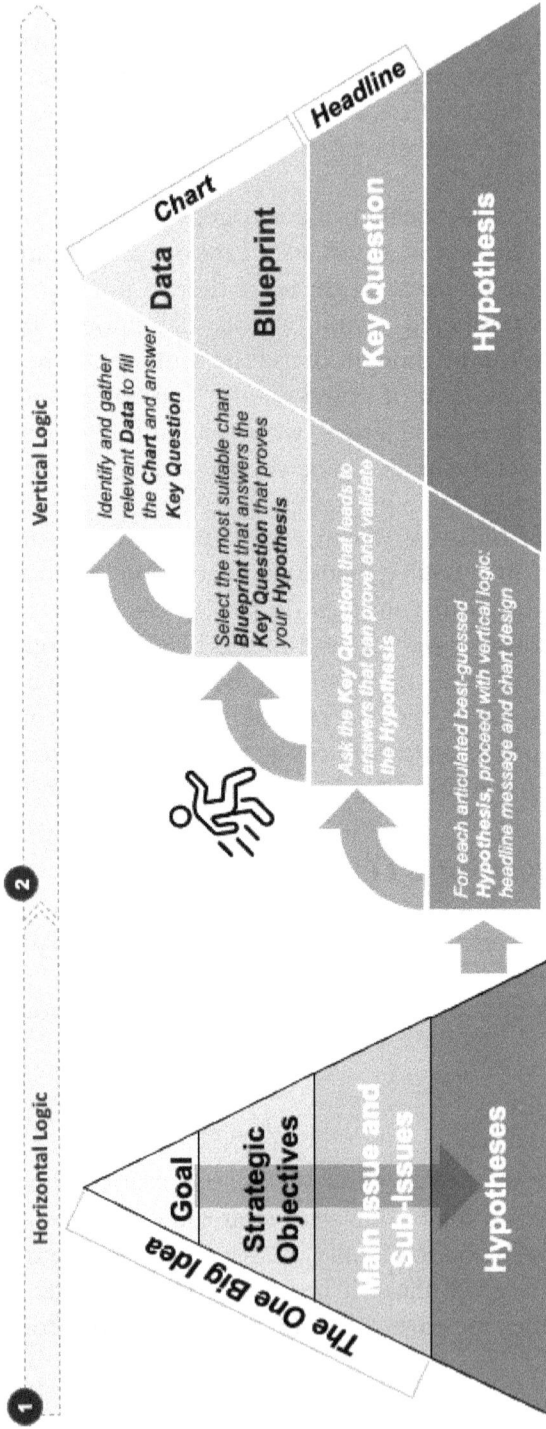

**Horizontal Logic**

**Vertical Logic**

**The One Big Idea**

- Goal
- Strategic Objectives
- Main Issue and Sub-Issues
- Hypotheses

**Chart**

**Headline**

- Data
- Blueprint
- Key Question
- Hypothesis

Identify and gather relevant **Data** to fill the **Chart** and answer **Key Question**

Select the most suitable chart **Blueprint** that answers the **Key Question** that proves your **Hypothesis**

Ask the **Key Question** that leads to answers that can prove and validate the **Hypothesis**

For each articulated best-guessed **Hypothesis**, proceed with vertical logic: headline message and chart design

This is why horizontal logic is considered a prerequisite for VL and needs to come before it. Horizontal logic is what defines your analysis and the data required to back up the claims and messages of your story.

## *Elements in a Slide and a Chart*

A chart is made up of several elements. These elements can be referred to in a variety of different ways. As such, different firms may give them different names during on-the-job training. Regardless of how these elements are named, though, the number of elements making up charts will always be uniform.

| R# | Element | Visible | Must | Definition and Purpose | Design Rule |
|---|---|---|---|---|---|
| 1 | Headline | ✓ | ✓ | The main message on a slide extracted from the story | No more than 2 lines |
| 2 | Title | ✓ | ✓ | The description of a graphic or a chart within a slide | Consistent location across all slides |
| 3 | Chart Area | | ✓ | The area in which the chart should fit and not exceed the margins allocated to it - as set by the guides set by this template | The area isn't marked but should be consistent throughout: Its borders should have the same distance from the top and the left |
| 4 | Ghost | ✓ | | A text or graphic illustrating position in a story within the story | Can be a text, image or both and goes between the header and chart area |
| 5 | Label | ✓ | | Description of highlights and visual cues | Location should be consistent across the entire deck |
| 6 | Ribbon | ✓ | | Ribbons can be used to highlight an attribute or something on the entire slide. Can also be used for disclaimer (see below) | It needs to go above the header |
| 7 | Disclaimer | ✓ | | Used to clarify and delimit the scope of proof in the chart that is used to support the headline claim or message | Disclaimers can refer to a specific chart or the entire chart area. Multiple disclaimers can be added |
| 8 | Callout | ✓ | | A callout is a text or a visual cue designed to stress or highlight a particular message in the chart | Avoid overcrowding the slide |
| 9 | Page # | ✓ | ✓ | Is a must on every page except the first page | Format and location should be consistent throughout |

| 10 | Source | ✓ | ✓ | The source of the chart or the data in the chart should go on every slide. Every chart should have a source, i.e., if you have 2 charts from different sources, include the 2 sources | Consistent location, font, size, format at the bottom of the page but above the footer if there is one. Separate different sources with semicolons |
|----|--------|---|---|-----------------------------------------------------------------------------------------------------------------------------------------------------------------------------------------|-------------------------------------------------------------------------------------------------------------------------------------------------------------------------|
| 11 | Footnote | ✓ | | This is a useful field to explain or justify specific elements, figures, analysis, etc. in the chart area | Use only when clarification or qualification is needed. Usually comes before the source field listed in numerical order and use 10pt font or less |
| 12 | Note | ✓ | | Unlike the footnote, the note typically refers to the entire chart and not specific elements within the chart. | Comes after the numbered footnote if there are any |
| 14 | Footer | ✓ | | A footer is only needed for branding, or adding disclaimers, copyright and confidentiality marks | It's recommended but may crowd out the slide |
| 15 | Margins | | ✓ | The position and the space and margins between the elements in the chart including the headline, the title, and the source/note field | Positions of titles, internal charts (i.e., for a single chart or multiple charts), margins, spacing between elements and the spacing from the top and the left should be consistent throughout |

The charts above provide you with the names and definitions of the most common chart elements. Where you place these elements in a chart isn't exactly set in stone because the only thing that matters when it comes to these elements is consistency. If you have remained consistent across the board, then the template you're using should be uniform in placement, color, fonts, and image sizes throughout your entire presentation.

The reference numbers (R#s) in the tables are visualized and illustrated on individual slides in the examples that follow.

The "Visible" column in this visual indicates whether a specific element or rule is about a visible element or not. The "Must" column, on the other hand, denotes whether it should apply to every single content slide in a deck.

## *The Relationship Between the Elements on an Individual Slide*

One of the ROTs of STC is that you should only have one message per slide in VL, if you recall.

Contrary to what most people think, presenting two ideas on a single slide takes the exact same amount of time as presenting them on multiple slides takes. If you are not constrained by the number of slides you can use—which you should not be—you can certainly split your messages into different slides. In doing so, you can keep from cramming multiple messages and charts into a single slide. Doing this will generate fewer questions and make your overall message easier for your audience to digest.

Whatever claim you're making or insight you're sharing should be explained in a one, or two-line, single sentence in your headline. That claim or insight should be fully supported by your data and the analysis that can be found in the chart that is located in the body of the slide.

There's a symbiotic and almost recursive relationship between the headline message and the body of the slide. This is because the body of a slide supports the message and the message, dictates what the body should contain. Something mentioned in the headline shouldn't be missing from the body. Likewise, something mentioned in the body shouldn't be absent from the headline.

1. Headline Area (Fits maximum 2 lines aligned to the left)

2. Chart Title Area (Can be split up into 2, 3 or 4 titles one for each chart, can be centered or left-aligned)

3. Chart Area

1. xxxx 2. xxxx 3. xxxx
Note: Use font B
Source: Include a source for every chart that you use. Separate each source with a semicolon followed by two spaces

Confidential And Proprietary

Business Model Hackers®

© 2023 ACME Inc All Rights Reserved

That being said, slides aren't exactly chicken and egg situations. Your message always comes before your chart is selected, as mentioned

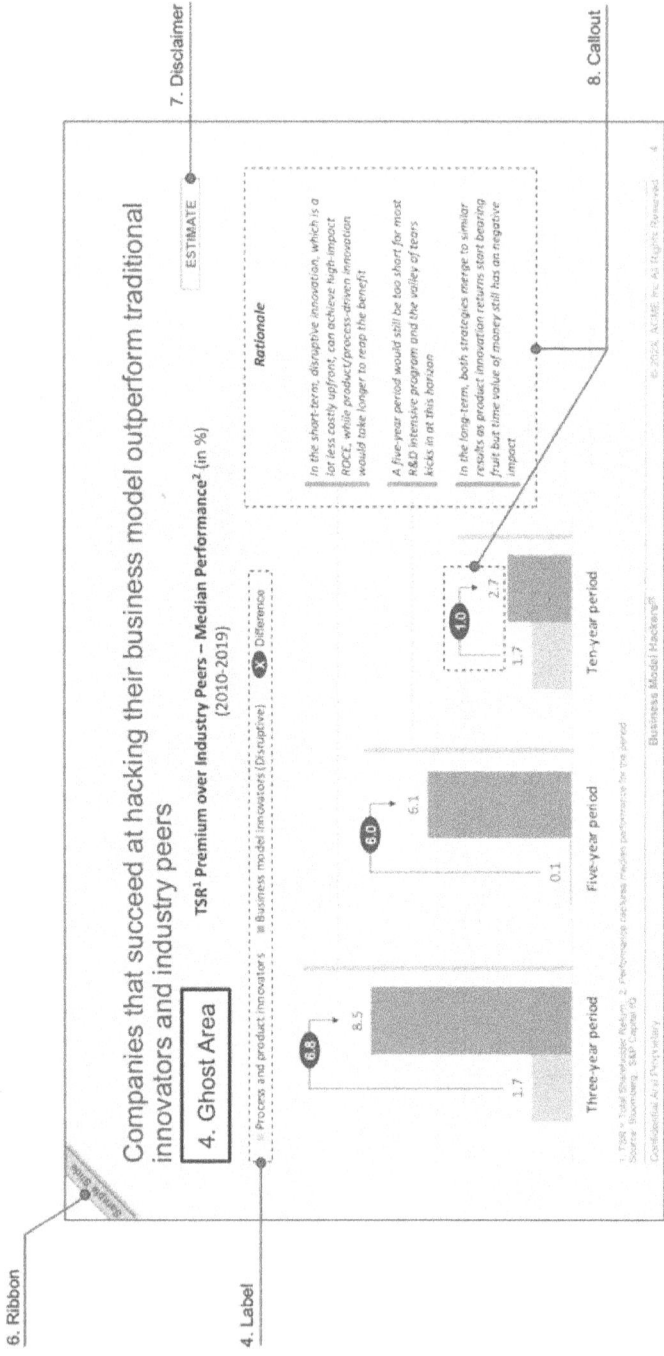

**Companies that succeed at hacking their business model outperform traditional innovators and industry peers**

TSR[1] Premium over Industry Peers – Median Performance[2] (in %)
(2010-2019)

Legend: ▨ Process and product innovators | ▨ Business model innovators (Disruptive) | ✕ Difference

**Rationale**

*In the short-term, disruptive innovation, which is a lot less costly upfront, can achieve high-impact ROCE, while product/process-driven innovation would take longer to reap the benefit*

*A five-year period would still be too short for most R&D intensive program and the valley of tears kicks in at this horizon*

*In the long-term, both strategies merge to similar results as product innovation returns start bearing fruit but time value of money still has an negative impact*

ESTIMATE

Three-year period: 6.8, 8.5, 1.7
Five-year period: 6.0, 6.1, 0.1
Ten-year period: 1.0, 2.7, 1.7

1 TSR = Total Shareholder Return. 2 Performance reflects median performance for the period
Source: Bloomberg; S&P Capital IQ
Confidential And Proprietary

Business Model Hacking[3]

© 2024, ACME Inc. All Rights Reserved    4

Labels: 7. Disclaimer · 8. Callout · 6. Ribbon · 4. Label · 4. Ghost Area previously, and determines the chart or visual that will be required to support it.

One ROT you should keep in mind is that you should be able to accurately guess what your headline says if you were to hide the headline and view the chart by itself. For the record, the reverse of this statement is also true.

## *The Look, Feel and Template Design*

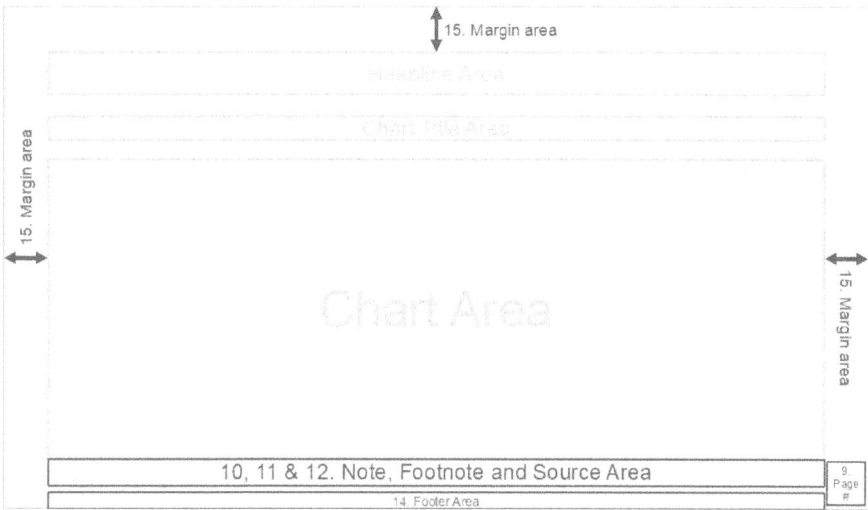

Having a template to work from is something that can ensure that all the chart rules remain uniform across your deck.

For reference, a template is a predesigned blueprint for your slides that contains your branding, fonts, layout, colors, and the like. In PowerPoint, templates can be viewed under Slidemaster.

Templates are pre-saved for frequent use and quick access. You can create custom templates, store them, reuse them, and easily share them with others.

You can get fancy with templates too and add different layouts and additional elements to them (Asión-Suñer & López-Forniés, 2021). If you want to gain access to the most commonly adopted templates available on the market and the ones that are used by MBBs, you can take a look at them in our toolkit.

An ROT to keep in mind is that the fewer pages in your templates, the better and the more consistent your deck's "look and feel" will be.

One rule you can never go wrong with is that less is more (Lidwell et al., 2010). As such, the focus of STC should be put on something other than fancy layouts and colorful slides. Instead, it should be on insights, story, and on backing your claim with absolute credibility. Focusing on these things first and foremost always works like a charm.

At least it has for me. Granted, some situations may call for more colorful decks and images, such as in designing storyboards, concepts, and illustrations. But the possibility of the very occasional exception doesn't negate the fact that your story always needs to be backed up by analytics and logic. Given that, your slides always need to be simple and get straight to the point, even if your analytical model and data analysis require more complex modeling.

The core concept here relies on simplifying your findings into something that any kind of audience can instantly absorb through vertical logic and that can be presented to them with just a handful of templates.

The must-have templates in any presentation are the title slide, the table of contents, the blank page, and the single-page chart. You can maybe include the double or triple chart page templates in this list as well.

Over time your library of reusable charts will grow as you keep building more presentations. What that growth will look like will depend on what your business and profession are. Soon you'll start pulling charts from other presentations and customizing their headlines. You'll also start customizing those charts, so that they fit in well with their new homes, instead of starting charts from scratch with every presentation.

Whether you are presenting virtually, onscreen, in a hall, or in a conference room, you can use the following chart table to guide your font size, assuming you're employing the standard Arial font.

Considering how much the work environment relies on technology these days, the fact that computer screens have become the standard presentation format shouldn't be a surprise. That being the case, should also be no surprise that smaller fonts are seldom legible during presentations.

| Best Viewing Distance by Screen Size | | | |
|---|---|---|---|
| Font Size (in pt.) | | Screen Width | |
| | | 3 meters | 4 meters |
| Can you read this? | 9 | 3 | 3 |
| Can you read this? | 10 | 3 | 4 |
| Can you read this? | 12 | 4 | 5 |
| Can you read this? | 14 | 5 | 6 |
| Can you read this? | 16 | 6 | 7 |
| Can you read this? | 18 | 7 | 9 |
| Can you read this? | 20 | 10 | 15 |
| Can you read this? | 22 | 12 | 17 |
| **Can you read this?** | 24 | 15 | 20 |
| Can you read this? | 28 | 17 | 24 |
| Can you read this? | 32 | 20 | 26 |

Despite this, I would recommend that you use 10 pt. fonts in all your slides, except for their Footer, Source, and Note sections. As for colors… You might think this is of secondary importance, but you would be wrong. The color you use in your presentation will be linked to your brand, after all. If you find that you struggle with colors, you can send your template to a professional designer. They can then develop your sample slides based on a chosen color palette. Alternatively, you can build your own color palette using the colors of the main logo you've been provided with and by applying different shades to them throughout the presentation.

If you have a single-color logo or brand to work with and insist on using more colors, a good strategy would be to find matching color combinations through a simple online search. If, on the other hand, you're working for a client who prefers adopting internal branding to the presentation, you should combine black and gray shades with your slides' main brand colors.

## *The Four Types of Charts*

There are two kinds of slides that are usually included in every deck. These are "fillers and bloaters" and "appendices and backup slides."

Fillers are the types of slides that include the Table of Contents (ToC), agendas (see example below), separators, or chapter slides. I refer to separators as using their secondary name, bloaters, because I'm usually not a big fan of them. Bloaters, however, are a necessary evil, especially if you're dealing with a big report made up of several phases of work streams, really long presentations, or workshop presentations.

Despite this, though, you should remember the ROT that the best stories flow flawlessly from beginning to end without the need for bloaters. If your story doesn't flow properly without separators, you may want to revisit your storyline.

Meanwhile, appendices are those slides that aren't necessarily a part of the story but could be referred to as backup materials or analyses supporting the specific messages that the main story gives. They're the executive summary presentations and backup charts you have on hand, just in case you're asked for them or in case you need to demonstrate how you arrived at a specific claim you're making.

In addition to all these, there are also your approaches and your framework charts. These are your structural, skeleton charts. Structural charts form your foundation. They're the spine that holds most, and sometimes all, of the storylines together. In my estimation, though, these can be considered bloaters as well. This is because they can be counted as smart replacements for ToCs and agendas, which brings me to qualitative and quantitative charts.

Agenda

Part 0: Introduction

Part 1: What Is STC?

Part 2: Why the STC Method Will Live Forever

Part 3: The Anatomy of STC

Part 4: The Anatomy of Charts

Part 5: The Universal Framework

Part 6: Using the Toolkit for Chart Selection

Part 7: Horizontal Logic and Storylines

Rules Of Thumb

Qualitative and quantitative charts are the last category of charts that can be included in slides. These charts are what make up a story when they're used in a specific sequence.

## The Distinction Between Stories and Storytelling

Before we can venture into the world of VL design and development approach, we have to establish what the distinction between stories and storytelling is. Your deck and its HL are what you use to tell your story, as we have established. When you run through your deck, your story should flow in the way you want it to. At the same time, every chart contained in your deck should be a story contained within itself. The well-known metaphor, "a picture is worth a thousand words", therefore, feels oddly fitting for STC. If you're making a short presentation of 10–25 slides, for instance, every chart in that slide should be a self–contained story.

If your horizontal logic is your main plot, then your vertical logic can be considered your sub-narratives. They make up short stories, scenes, and moments that add details and value to your overall plot (Boyd et al., 2020). It's hard to express a full and complete story on a single slide in visual format if you are not doing a voice-over narration over it. This is why we have to distinguish between in-person presentations vs. executive summary decks accompanied by an in-person presentation.

Understandably, storytelling is much easier to do in the in-person or voiceover presentation format. This format literally gives you the opportunity to explain what each slide and your overall story is indicates and what message they're conveying. It reduces the risk that something will become lost in translation or difficult to grasp.

Compared to the in-person format, the deck–based storytelling without voiceover is much more difficult to do. This is because you have to make sure you communicate every message you mean to give clearly and concisely, but without missing anything when using this format. If you prepare a poorer deck, you risk things becoming lost in translation, without you there to provide voiceover explanations and clarifications.

## *The Universal Framework for Vertical Logic*

VL is the most straightforward concept you can learn in STC. You neither need any kind of special training nor need to read really long, thick books to master VL. All you really need to do is master some basic rules that you can quickly memorize or sum up on a handy cheat sheet, which you can use until this information becomes intuitive or second nature to you.

A core STC skill for you to make note of is how HL design is used in structured thinking. You can make use of several mental techniques and methods to structure your thinking abilities. One such technique is called the issue diagram. Alternatively referred to as the hypothesis-driven approach, the issue diagram has you formulate a hypothesis based on what issues or sub-issues you have on hand. You, of course, derive these issues and sub-issues from your strategic objectives and main goal.

Assuming you have come up with a hypothesis, your next step is to ask the right questions so you can prove that hypothesis to be correct. In the issue diagram, you can break down the VL into a simple Five-Step Universal Framework (5SUF) that can work for any chart.

As you'll recall, the horizontal logic leads to individual claims or messages. In the case of the issue diagram, your starting point is a set of hypotheses. Why hypothesis? Because you have yet to prove them, and you need to build the verticals you need so that you can demonstrate each of them.

Now that we have established all this, let's take a closer look at how this universal framework works. As you can see in the given issue diagram, the vertical design is actually an iterative step. The first step in the framework is the key question. The key question is a fixed point. The second and fifth steps of the framework, on the other hand, are not fixed. They're iterative as well. Before delving deeper into each and every one of these steps, let's clarify how 5SUF works and to differentiate between what quantitative and qualitative charts are:

5SUF starts with the hypothesis. From there you develop your key question. This is step 1. Next, you answer that question in your headline, which helps you to develop your blueprint. This is step 2. You can now use your blueprint as a guide for your analysis, which will in turn narrow down your data requirements for you. That is step 3. The analysis informs your data needs, which is step 4. Finally, the narrowed down and chosen data will be displayed in a compelling visual in the form of a chart, which is step 5.

Moving on, quantitative charts—the primary focus of this chapter—are data and number-driven charts. Qualitative charts, on the other hand, are visuals that are primarily text driven.

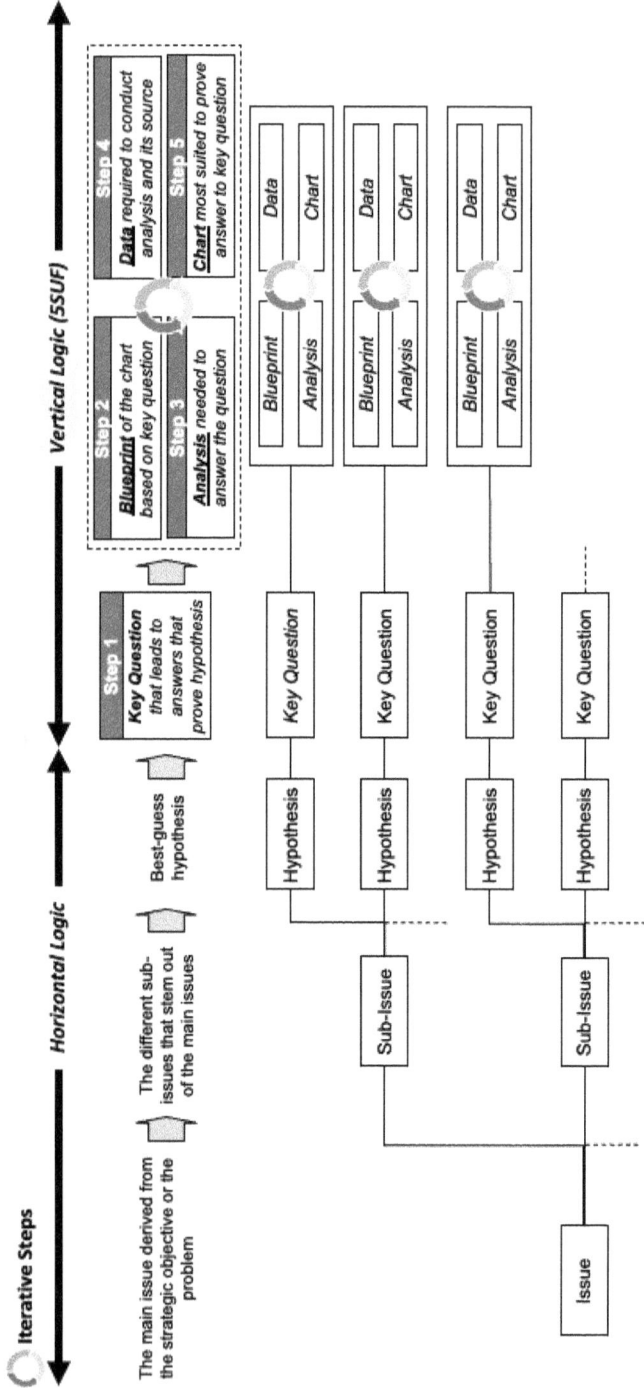

Iterative Steps

Horizontal Logic — Vertical Logic (SSUF)

The main issue derived from the strategic objective or the problem

The different sub-issues that stem out of the main issues

Best-guess hypothesis

**Step 1**
Key Question that leads to answers that prove hypothesis

**Step 2**
Blueprint of the chart based on key question

**Step 3**
Analysis needed to answer the question

**Step 4**
Data required to conduct analysis and its source

**Step 5**
Chart most suited to prove answer to key question

Issue

Sub-Issue

Sub-Issue

Hypothesis

Hypothesis

Hypothesis

Hypothesis

Key Question

Key Question

Key Question

Key Question

Blueprint | Analysis | Data | Chart

Blueprint | Analysis | Data | Chart

Blueprint | Analysis | Data | Chart

## *Quantitative or Data-Driven Charts*

One thing I always tell my trainees and apprentices is that you should only go for qualitative charts if you cannot find the numbers and quantitative data to back up claims or messages. You should always aim to drive your story forward using quants and have as many quantitative charts on hand as possible to back up your full story. This is because arguing with numbers is harder to do than arguing with more abstract data. The numbers that drive a given chart and thus drive a specific story often says it all, so long as you pull the right numbers to support your claims and headers.

Claims that are supported by factual data and solid reasoning are more compelling than claims supported by subject matter experts, focus groups, or other qualitative evidence. Therefore, an ROT you need to bear in mind is to let the quants drive your story and prioritize quantitative slides over qualitative slides.

That being said, some people take this advice a little too far and go overboard with data visualization. Doing so can be tempting, as it will give you an opportunity to show off your technical skills by choosing the most complex visuals available. This, however, will be counterproductive to what you want to do. What you should be doing instead is prioritizing simple charts as discussed earlier, since they will always be accepted as proof.

## *Step 1: The Key Question*

There is a quote that's often attributed to Albert Einstein: "If I have an hour to solve a problem, and my life depended on it, I will spend the first 55 minutes determining the proper questions to ask, for once I know the questions, I could solve the problem in less than 5 minutes"

Whether this quote was really said by Einstein is unimportant. What is important is that it demonstrates how solving problems is all about asking the right questions. In our case, where vertical logic is concerned, the problem we're trying to solve is how to support the claims we're making and the messages we're giving in our slide's headline.

| Issue: Sales and profits are declining | |
| --- | --- |
| **Sub-Issue 1: The company is losing market share to new entrants and competitors** | |
| **Hypothesis** | **Question** |
| The technology adopted in the company's product is old dated and undesired by customers | Is a new or emerging technology replacing the old one? |
| Growing attrition of users who are switching to substitute products | How many customers switched over the past three years? Where did customers purchase or switch to? |
| Product prices of like-for-like features are too high | How do the product prices compare to competitors' products for the same features? |
| Users are increasingly dissatisfied with the product | Are users unhappy with our product? |
| A new product launch is required to retain customers | What happens to our sales or financials if we don't launch a new product and the current sales trend continues? |
| A new state-of-the-art technology will be required to remain competitive | If the company copies a competitor, wouldn't it be just playing catch up? |
| Acquiring a new company or startup in a new emerging technology space is one of the best and quickest ways out of the vicious circle | What are the options to get out of this vicious cycle and how do these options compare to each other? |

Accomplishing this requires proving our hypothesis. We don't need to worry about how our hypothesis came into being at this stage. Instead, we need to focus on the questions we need to ask so that we can either prove or disprove our hypothesis.

When recruiting new people to our industry, we try to find those candidates who are able to ask necessary questions like this. When we recruit or interview strategy consultants from top MBA schools, we screen for candidates who demonstrate an ability to think strategically. We do so by checking if they can come up with the right questions when solving problems in case interviews.

The importance of structured thinking may not be blatantly obvious at first. This doesn't change the fact, however, that every aspect of work and life revolves around our ability to effectively structure our thoughts, plans, and data. This kind of thinking requires quite a bit of discipline because, as humans, we intuitively think about many different things simultaneously.

The final step in the process is data analysis and selection, which can be guided by the answers you give to the questions you ask, designed as they are to prove the hypotheses you've made. Asking the right questions, then, is a core skill you need to learn across all levels, starting with HL, meaning the entire story, down to all the individual messages that make up the story, meaning VL.

To show what this process might look like, let's take a hypothetical example:

Suppose that your issue is that your company is losing market share to new entrants and competitors. You can come up with any number of hypotheses explaining this situation and even more questions that could be asked about them, as exemplified in the chart below.

# Quantitative Chart Selection Universal Equation

## Chart Type = F(T, V, MA)

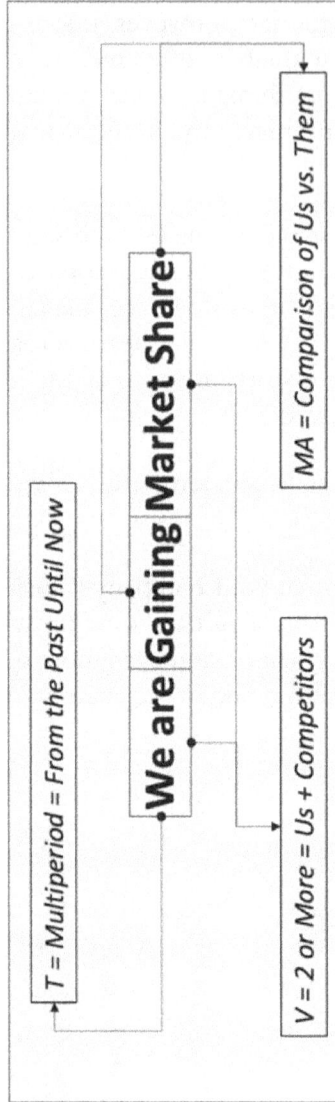

T = Multiperiod = From the Past Until Now

**•We are Gaining Market Share•**

V = 2 or More = Us + Competitors

MA = Comparison of Us vs. Them

| Symbol | Variable | Explanation |
|---|---|---|
| T | Time Variant | Is there a time factor, meaning is it static in time or trending over time? The answer is yes or no |
| V | Number of Variables | Is it a single or multiple variables? The answer is the number of variable, i.e., 1, 2 3 or more |
| MA | Message Attribute | What is the insight that needs to be shown? i.e., Is it a comparison, breakdown, relationship, distribution or frequency? |

## Step 2: Blueprint

Once you have asked your questions, you can move on to step 2, which is the blueprint phase. The blueprint is the methodology you use to select the type of chart you'll need in the early stages of your work. This is the most important step of 5SUF as it sets the scope and the focus of both the analysis and data mining processes.

You might be tempted to use qualitative charts at this stage, as always. But you should remember that numbers always tell a better story than verbiage does. So, unless you work in the legal field, you should always be able to rely on quantitative charts and numbers to back up most of your claims.

As an ROT, let the quants drive your story, and always prioritize quantitative slides over qualitative slides. Remember that numbers speak louder than words.

When selecting which chart or data to use to illustrate or back the claim you're making, start with the claim or the eventual headline that you derived from the hypothesis, as well as the question you asked in Step 1 of this process.

Use the following formula when selecting your chart:

Chart Type = F (T, V, MA) where

T= Time, V = Variables, and MA = Message Attribute

You determine T (time) by establishing whether the issue at hand has a time factor. It either does or it doesn't. You determine the number of variables by establishing whether you're looking at a single variable or multiple variables. You'll be able to find the answer to that question in the form of the number of variables. Finally, you determine the message attribute by asking yourself what the insight that needs to be shown is.

Fundamentally speaking, there are five attributes you need to consider when working to identify which type of quantitative chart you are to use. These are hierarchy, breakdown, frequency, relationship, and comparison.

This formula is referred to in STC as TVMA. The secret to identifying which chart to use lies in deciphering the message and determining the value of the three variables in the equation.

As an example, let's say that your message or claim is "We are gaining market share."

This message implies that we are comparing yourselves to our competitors over a specific period of time, stretching from the past till today. This means that:

T = There is a time component to this equation.

Variables = "Us" + competitors, which means that we have two variables.

Message Attribute = Comparison, seeing as our market share is being compared to that of our competitors.

There are a total of 16 families of charts that you can use to visualize almost every quantitative scenario you can imagine.

## The 16 Families of Charts

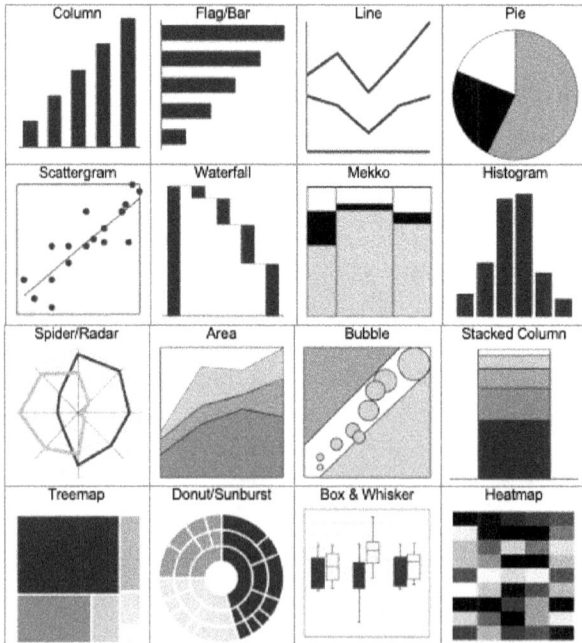

Using both "icons" and texts, this visual clearly shows how the answers you uncover using your equation can be plotted into your chart blueprint. So, if you were to use the flowchart for our market share example, you would begin your work by identifying your message attributes first. In this case, this is "comparison." Then you'd move on to time, which is changing as opposed to static. Finally, you'd identify your variables, which you have two of. You can then use this information to determine which of the chart options is the best fit for you. In this case, the flag chart, bar chart, or the line chart would be viable options for you.

### Signal/Trigger Words

When crafting your message or analyzing your data, be on the lookout for certain words or phrases that can guide you towards the most appropriate chart category. These signal words serve as indicators of the type of comparison or relationship you want to illustrate. Below, you'll find a list of examples signal words for each Message Attribute:

### Comparison:
- Compare, contrast, differentiate
- Higher, lower, bigger, smaller, greater, lesser
- Increase, decrease, grow, shrink, expand, contract
- Improve, decline, enhance, deteriorate
- Outperform, underperform, surpass, lag behind
- Benchmark, baseline, standard, norm

### Relationship:
- Relate, correlate, associate
- Influence, impact, affect, shape, determine
- Cause, effect, result, consequence
- Depend on, rely on, link to, tie to
- Connect, interact, interplay
- Trend, pattern, trajectory

### Frequency or Distribution:
- Frequency, occurrences, incidence, prevalence
- Distribution, spread, dispersion, allocation
- Range, spectrum, scale, scope
- Allocate, divide, apportion, partition
- Cluster, group, gather, aggregate
- Classify, categorize, segment, sort

### Breakdown or Composition:

- Break down, split, separate, dissect
- Compose, make up, constitute, form
- Percentage, proportion, fraction, ratio
- Share, portion, piece, slice
- Segment, part, component, element
- Contribute, account for, attribute to

### Hierarchy:

- Rank, order, position, place
- Top, bottom, upper, lower
- High, medium, low, mid
- Superior, inferior, above, below
- Priority, importance, significance, urgency
- Sequence, progression, series, succession

It's important to note that these signal words are not mutually exclusive and collectively exhaustive (MECE). Some words might fit into multiple categories depending on the context and the specific message you're trying to convey. For example, "increase" could indicate a Comparison (comparing two values) or a Relationship (showing how one variable increases with another).

Use these signal words as a starting point to guide your thinking, but always consider the overall context of your message and the data you're working with. As you become more familiar with these categories and signal words, you'll develop a keen sense of which chart type will most effectively communicate your story.

Remember, the goal is to choose the chart that best aligns with your message and the comparison or relationship you want to highlight. By paying attention to these signal words and understanding the different chart categories, you'll be well on your way to creating compelling and impactful visuals that bring your data to life.

Additionally, you can use the table/matrix format to determine which type of chart is best suited for which type of equation component. As you gain experience though, you will find that you've started committing this information to memory, until it becomes intuitive at last

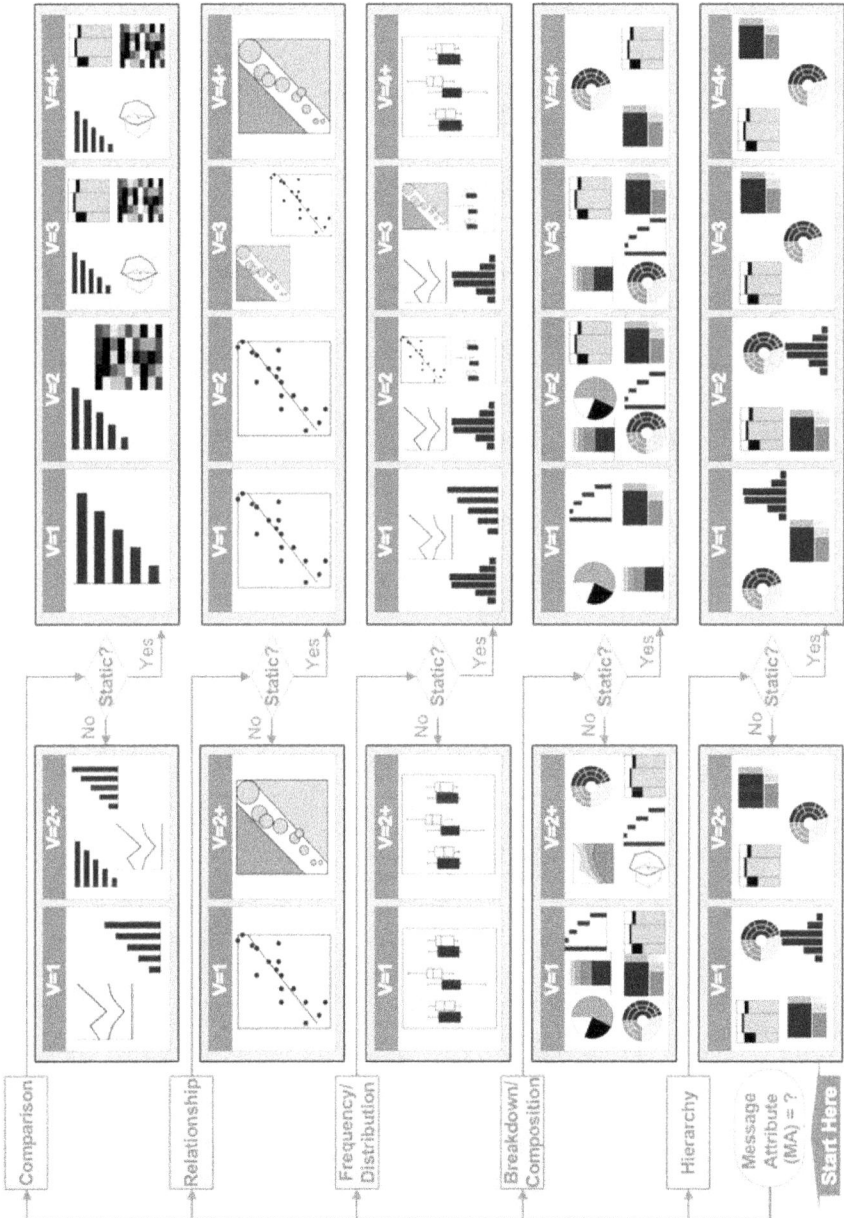

| TV► / MA▼ | Static: Single Point in Time | | | | Trending: Changing over Time | |
| --- | --- | --- | --- | --- | --- | --- |
| | **1 Variable** | **2 Variable** | **3 Variable** | **4+ Variable** | **1 Variable** | **2+ Variables** |
| **Comparison** | Flag | Multi flags, Heatmap | Mekko, Multi flags, Spider/Radar, Heatmap | Mekko, Multi flags, Spider/Radar, Heatmap | Line, Column | Line, Column, Flag |
| **Relationship** | Scattergram | Scattergram | Bubble, Surface/3D Scattergram | 3D Bubble | Scattergram | Bubble |
| **Frequency/ Distribution** | Histogram, Single line, Column | Multi histograms, Multi line, Scattergram, Box and Whisker | Multi histograms, Multi line, Bubble, Box and Whisker | Box and Whisker | Box and Whisker | Box and Whisker |
| **Breakdown/ Composition** | Pie, Waterfall, Stacked Column, Treemap | Multi-stacked column, Multi pie, Mekko, Donut/Sunburst, Waterfall, Treemap | Multi-stacked column, Mekko, Donut/Sunburst, Waterfall, Treemap | Donut/Sunburst, Treemap, Mekko | Multi pie, Stacked Column, Waterfall, Donut/Sunburst, Treemap, Mekko | Area, Donut/Sunburst, Multi Radar, Waterfall, Mekko |
| **Hierarchy** | Donut/Sunburst, Treemap, Histogram | Donut/Sunburst, Treemap, Mekko, Multi histogram | Donut/Sunburst, Treemap, Mekko | Donut/Sunburst, Treemap, Mekko | Donut/Sunburst, Treemap, Mekko, Multi histogram | Donut/Sunburst, Treemap, Mekko |

## Step 3: Analysis

Analysis is the step you use to ascertain a hypothesis' validity by doing data mining and finalizing your chart. The analysis is a kind of computation that you may need to do on your data set. Doing analysis can require something as simple as creating an excel sheet where datasets are reorganized so that they can fit into a chart. It may also be as complex as working with dynamic or optimized models.

Complex modeling is something you typically only have to do if you actually want to do it. For instance, it can measure the impact of a particular initiative or recommendation. It can also test out different assumptions and possible scenarios and can be used when you want to optimize a goal like increasing company valuation or profitability based on varying multiple assumptions or input.

On the whole, measuring the impact your recommendations have is usually a good idea. However, seeing as most analysis is usually made in the form of an assessment, going over three common sense methods often employed in analysis is a good idea. If you want to learn even more about these methods, then all you need to do is a simple web search. It's important, however, that you remember how most of the analytical work in the STC framework relies on common sense, rather than your ability to use some complex, magic formula.

### The First Method: Fermi Thinking

Fermi thinking is an estimation technique that gets its name from the physicist Enrico Fermi. It has been adopted to solve extreme problems that cannot be easily resolved through mathematical or scientific means.

Fermi thinking allows you to arrive at a solution by adopting answers and using them as orders of magnitude estimates. Put in layman's terms, Fermi thinking has you break down problems into smaller chunks and then divide those chunks into two piles you'll label as "known" and "unknown". This allows you to become aware of what you know and what you need to learn. You can then use what you know to figure out what you need to learn (Chakraborty, 2020).

Fermi thinking, then, is a quick and simple way of developing a frame of reference for what you might expect your answer to be. As an ROT, and in the context of data validation, you should apply Fermi thinking where applicable, to check if the data at hand is within a reasonable range.

### *The Second Method: Data Triangulation*

Triangulation is a term that has been derived from the Latin word "Triangulum". Within the context of data mining, it's the metaphorical reference you use to validate how accurate data points are. You do this by examining the data from different points and then cross checking them with various data sources.

### *The Third Method: The 80/20 Rule or the Pareto Principle*

You might already be family with the 80/20 rule. If you're not, then I would suggest doing a quick search online as this is an important tool to have in your arsenal. So much so, that it's also an ROT for when you're analyzing independent data sets like assessing how product defects are caused by 20% of the issues in the production line. The 80/20 rule basically says that 80% of the problems in any given scenario are derived from 20% of the causes (Tardi, 2022). By virtue of its logic, this rule can be used to figure out and focus on what matters most in a given situation (Grigorieva, 2015). That might be fixing 20% of the problems at hand, since by fixing them, you'd be eliminating 80% of the issues you're having anyway.

## *Step 4: Data*

The truth is you might not always have to use data for every single presentation. But since quantitative charts are more powerful and effective in proving claims than verbiage is, using data won't ever hurt.

Unless you're working on a legal-based deck, numbers always speak louder than words. That said, having legal proofs and claims sprinkled throughout your slides can actually benefit your STC. This is because they increase engagement and structuring, as you'll see later in the qualitative charts chapter.

For now, the key question to consider about quants is whether you have the numbers you need to back your claims and when you should put them together. Your numbers should ultimately hold and validate your claim and hypothesis. If they don't, then you may need to re-evaluate them. If they do, then you need to assess your data using the following criteria:

- source credibility/authority
- relevancy
- accuracy
- completeness
- recency

Having several ways to back up a claim is always a good idea. Given that, repeating steps 2 and 3 of the Universal Framework once you've finished collecting your data can only reinforce your choice of a blueprint. It can also boost the confidence you have in your proof and evidence before you move on to the last step, which is to finalize your chart. Having several data sources will also enable you to gather the evidence you need to prove your claim.

As you go about this process, though, you should always remember that acquiring data can sometimes be tricky. You should also remember that it's not about how much data you've been able to gather. It's all about the process of identification you've used and the quality of the data you've obtained.

## *Step 5: Chart*

Step five is when the old saying "a picture is worth a thousand words" finally becomes realized. This saying couldn't be truer for vertical logic. Hence, the chart you pick should reflect your message without saying it in words.

An ROT to remember here is that the fewer words you use to make your claim or message come to light, the better your chart will be.

Once you're done mining and analyzing data, you'll have to finalize the visualization of the chart you've selected. This is the easiest part of the entire process because it's the part where everything falls into place.

It's worth mentioning, though, that the TVMA blueprint may lead you to multiple chart blueprint options. Sometimes, it may lead you to as many as six different possibilities. That is perfectly alright. If and when you end up with multiple chart options, you'll be able to choose the most suitable one based on which data and data sets are available to you now or as you iterate. Nonetheless, to avoid endless iterations, we can cut these into three final substeps:

**Examples of Quantitative Charts**

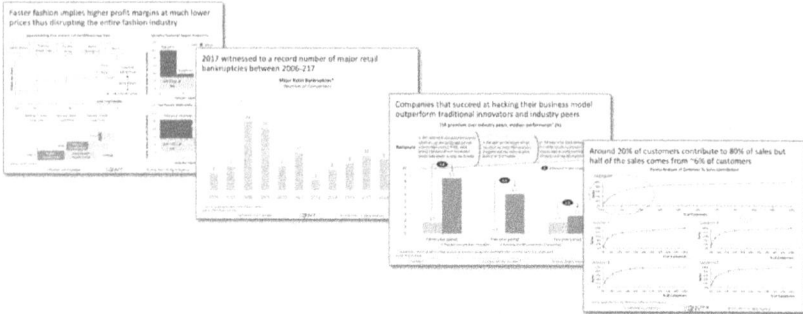

### Substep 1: Finalize and Crystalize your headline message

- Revisit the message or main point you want to convey with your chart.
- Ensure that your message is clear, specific, and focused on the key takeaway or aspect of the data you want to emphasize.

### Substep 2: Finalize your blueprint

- Based on your message, determine which of the five basic comparisons, component, relationship, frequency/distribution, breakdown, and hierarchy.
- To do so, extract the trigger words and clues discussed earlier to guide you in identifying the appropriate attribute.

### Substep 3: Select the most suitable chart

- Refer to the chart selection table that matches your identified message attribute with one of the sixteen families of charts.
- Choose the chart that best aligns with your message and the attribute you identified.

Another ROT to remember then is that you should sketch out multiple versions of the data and pick the visual that is most suited to backing your claim.

As a final note, oftentimes you might have to combine several charts under a single headline, message, or claim. Once you've mastered how the family of 16 charts are used, you'll be able to mix and match different ones without issue and use chart combinations easily.

## *Visual Cues*

What would happen if you were to turn a flashlight on while outside, on a sunny day? The flashlight would have little to no effect and would

**The Top Three Reasons Why M&A Deals Fail**

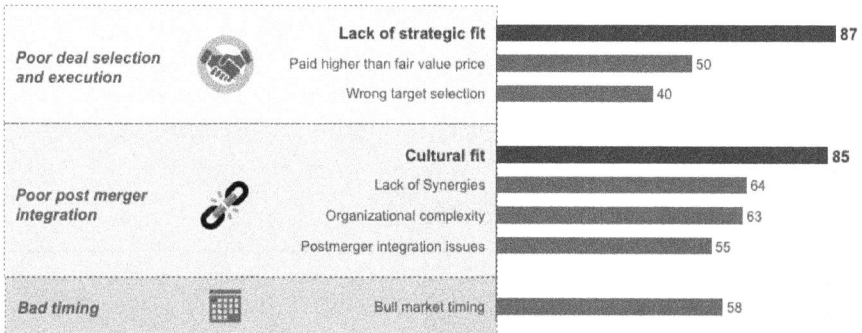

| | | Value |
|---|---|---|
| **Poor deal selection and execution** | Lack of strategic fit | 87 |
| | Paid higher than fair value price | 50 |
| | Wrong target selection | 40 |
| **Poor post merger integration** | Cultural fit | 85 |
| | Lack of Synergies | 64 |
| | Organizational complexity | 63 |
| | Postmerger integration issues | 55 |
| **Bad timing** | Bull market timing | 58 |

basically be useless. Using chars that don't have any visual cues to focus the reader on the message they're trying to give would be the same as turning on a flashlight outside in broad daylight.

Visual cues offer several advantages when used in vertical logic and charts. Of these, three are particularly important within the context of STC. The first of these advantages is that visual cues, particularly when dealing with charts that aren't easy to interpret, help guide the reader's focus to the part of a chart that reinforces the message its headline is trying to give (Lidwell et al., 2010).

How can you use visual cues to accomplish this? You can do so by playing with font sizes, colors, and shades, as seen in the example below. You can also make use of icons, which have become quite popular in recent years. You shouldn't use icons excessively, though, seeing as this would reduce how effective they are as cues.

**The second most important advantage** that visual cues have to offer is the fact that they can serve as memory cues. Memorability is a very important in STC as you want your audience to remember the important points you make. In other words, you want high audience retention. Visual cues can be very memorable in and of themselves and thus increase retention (Stafford & Grimes, 2012). At the same time, they can get your audience to act by having them recall the things you presented later on.

**The third most important advantage is chart interpretation.** To see this advantage at work, consider the phrase "I didn't say he stole the cash." If you read this phrase without emphasizing any of the words, nothing about it will grab your listener's attention. It'll simply mean that you didn't steal anything. But if you were to stress the words "say" or "he" in the sentence, then you would be giving a whole other meaning, complete with the "wink, wink, nudge, nudge" kind of indication. Stressing different words in a sentence, by making them **_bold, underlined, or italicized,_** for instance, can achieve this same result in presentations and open up your charts to greater interpretation, as can be seen in the example below.

| Stress Word | Meaning | Read |
|---|---|---|
| I | She is trying to say that she is not the person who said the man stole the money. Somebody else said it. | _**I**_ didn't say he stole the cash. |
| didn't | She is trying to say that she is not the person who said the man stole the money. Somebody else said it. | I _**didn't**_ say he stole the cash. |
| say | It sounds like she wanted to suggest that the man stole the money. But she did not want to say it directly. | I didn't _**say**_ he stole the cash. |
| he | It's suggesting that someone else stole the money, not the man identified in the sentence. | I didn't say _**he**_ stole the cash. |
| stole | It might mean that the man just borrowed the money. Maybe he didn't steal it. | I didn't say he _**stole**_ the cash. |
| the | The speaker is suggesting that she is talking about some other money, not the specific money that is being discussed. | I didn't say he stole _**the**_ cash. |
| cash | The indication is that the man stole something else. For example, maybe he stole jewelry or some other valuables. | I didn't say he stole the _**cash**_. |

The key takeaway here is that the meaning of your chart can change based on what you choose to emphasize. The same goes for visual cues, in that how and where you use them can help you to emphasize different parts of a chart. Using certain kinds of colors or shades in a chart can add some life to your vision and enhance the meaning you're giving to it. Color should not, however, be used as a part of your core messaging. If a chart shows that one variable occurs more often than another, for instance, it can make the bars showing this for these variables' different colors. This would emphasize the difference between the bars, but the key message the bars give would be conveyed through the fact that they are of different heights.

When you're working on charts, keep in mind that some executives still prefer working on and with printed paper and don't always print out the presentations given to them in color. This means that you might work on a chart and emphasize certain parts with colors, but those emphases would be lost if the person you give your presentation to prints it out in black and white. Add to that the fact that there are 300 million people across the world who are colorblind and that 4% of the US population is colorblind, and relying on colors in charts becomes a decidedly poor idea (Clinton Eye Associates, n.d.; National Eye Institute, 2019).

## *Qualitative or Conceptual Charts*

Qualitative charts, sometimes referred to as conceptual charts, are used to visualize the dense text in a structured format. As mentioned previously, you can build a good storyline using a combination of mainly quantitative charts and a few supportive qualitative charts. The obvious exception to this rule, of course, is fields that rarely have to rely on numbers, like legal advisory.

A legal advisor would have to prepare a deck or presentation that may be entirely made of text content. If they're working on something like case studies, they'll have to rely less on numbers as such studies don't have a lot of quantitative charts but have plenty of qualitative ones.

So, how can you convert verbiage into slides without risking audience engagement?

The most recent version of Microsoft PowerPoint has attempted to build such visuals under the "SmartArt" option. However, the options provided here, unfortunately, lack the flexibility you need to have in freeform art like this. Of course, there are software tools and add-ons you can turn to add automated freeform versions of this art onto your deck. Personally, though, I have yet to find a tool that is flexible enough to amend the shapes I'm provided with into freeform ones.

If you have downloaded STCAI, then that means you now have access to 2,000 freeform qualitative charts. You can quickly select charts by looking them up in the cheat sheet and edit your chosen visual with your own data in minutes.

## Qualitative Chart Categories

In the meantime, you should know that there are several categories of qualitative charts. The visual below has grouped these together by name and provides you with the most common ones. The categories that qualitative charts fall into are:

- **The Text to Visual Metaphors:** These are the most common charts. They are simply variations of the same concept. You convert text into a concise and nice visual to keep the user engaged. These also include table charts.
- **The Conceptual Frameworks:** These are explanatory charts that illustrate intellectual concepts visually.
- **The Story Flow Framework:** These are simple, flowing charts used to guide the reader through a complex deck or a storyline within a storyline.

## The Text-to-Visual Metaphor Qualitative Chart

The purpose of a qualitative chart is to convert text and paragraphs into mentally engaging visuals (Winkielman & Cacioppo, 2001). Of course, there is some science, that is to say, a method to the madness of accomplishing this.

Unlike other qualitative charts, the number of metaphors that can be adopted in this type when visualizing a text are endless. The sky's the limit in terms of which visual you can adopt as a metaphor.

The most experienced presenters usually use the same kind of bullet point or picture slides when running through text presentations or slides. However, you should avoid bullet points at all costs in STC. Otherwise, you'll end up boring your audience to death and preventing engagement.

Instead of using bullet points, you can make use of different types of art and thus add a bit of liveliness to your deck. In doing so, you can increase engagement and content retention (Stafford & Grimes, 2012).

To simplify this process, I have summarized the most common and popular metaphors you can use. While the list is not exhaustive, it should be enough to inspire any kind of messaging. Thanks to this inspirational cheat sheet, you should never run out of ideas for visuals to keep your audience engaged.

It's important to emphasize, however, that the visual itself is less important than the structure your qualitative argument follows in a qualitative chart. The process of structuring and organizing your main messages is the key to solidifying your VL.

# Text-to-Visual Metaphor Examples

## Visual Illustration of Corporate Business Unit Portfolio Distribution

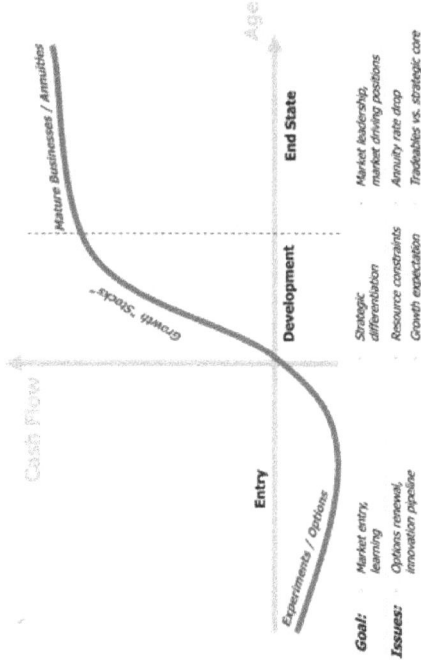

Cash Flow

Experiments / Options — Growth Stocks — Mature Businesses / Annuities

Age

**Entry**
Goal: Market entry, learning
Issues: Options renewal, innovation pipeline

**Development**
Strategic differentiation
Resource constraints
Growth expectation

**End State**
Market leadership, market driving positions
Annuity rate drop
Tradeables vs. strategic core

## Business Unit Portfolio Management Strategies in the Form of a Lookup Table or Cheat Sheet

| Portfolio Balance / Value Creation Lever | Options | Growth Stocks | Annuities |
|---|---|---|---|
| **Margin Plays (Market Discontinuities)** | Can we supercharge the upside by buying assets with significant optionality? Which businesses may grow faster than market? | In which sector and geographies can we accumulate assets at the bottom of the cycle? | Which markets will experience precipitous margin decline positions are tradeable? |
| **Strategy Plays (Competitive Discontinuities)** | Where are we experimenting to define the next strategic leadership play and which innovations will define future market leaders? | Which markets offer big bang opportunities from leveraging our strategies and where can we redefine the rules of the game in large markets? | Where are we the undisputed market leaders, able to set the terms of trade? |
| **Turnaround Plays (Operational Discontinuities)** | Can we enter somewhere advantageously by buying or turning a weakling around? | Where can we roll-up or fix weak competitors? | Which under-performing mature businesses offer the potential for market leadership and are there any mature business units worth fixing? |

So, with that in mind, how do you go about crafting your visuals? The first thing you need to do is to structure or organize your content before putting it into slides. To that end, ask yourself:

- What are my main themes and ideas? How many main themes or topics are there? What's the best way to group the content?
    - Grouping the content into themes enables you to determine the number of elements you will need in the visual.
- Do all the themes link to the main message or claim?
    - o If a theme or message does not support your headline or claim, then it should be removed.

## *Conceptual Framework Charts*

Conceptual framework charts are explanatory charts designed to simplify how an intellectual or complex concept is illustrated in an easy-to-grasp visual. This is often referred to as a framework.

The best way to explain conceptual charts is by example:

# Examples of Popular Conceptual Frameworks

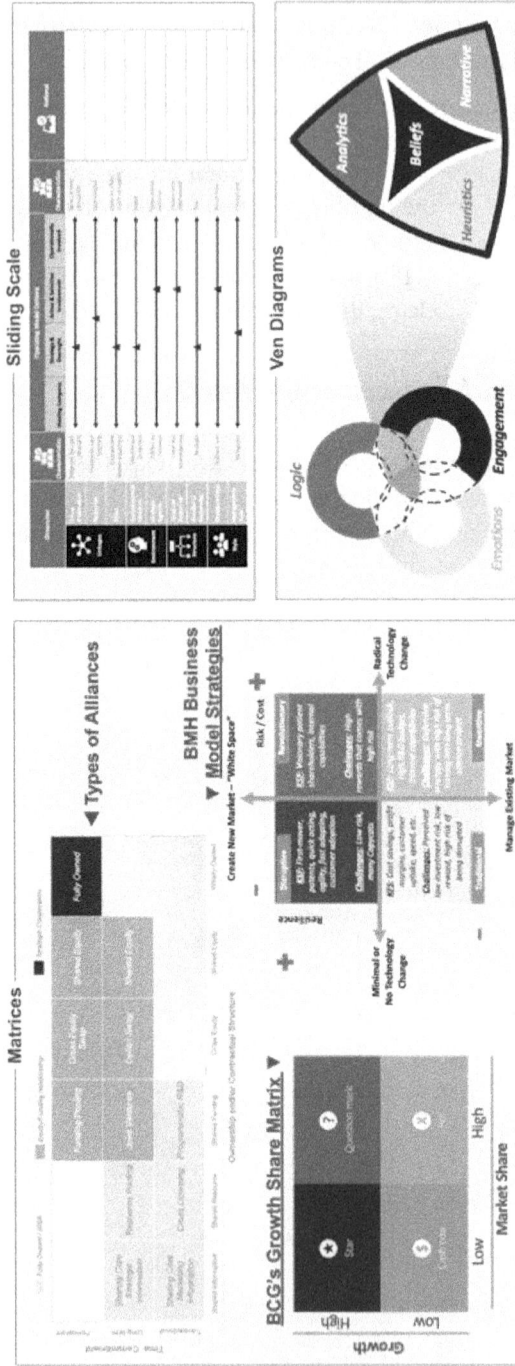

## *Matrices and Categorization*

If you've already come across the famous 2x2 Growth Share Matrix developed by Boston Consulting Group (BCG) in the 70s BCG, you know that the concept is simple. It's so popular that it is still in use and is frequently referred to in business to this very day. This shows how simplicity can leave a lasting impact and why this category of charts, if well developed, can leave a lasting impression on your audience and become a memorable experience for them.

Initiatives can be prioritized by adopting a three-step sequential and iterative process across four interdependent criteria

Initiatives Prioritization Framework

ILLUSTRATIVE

I Benefit

II Risk

III Priority

The BCG matrix has four quadrants (Hayes, 2022b):

- low growth, high share
- high growth, high share
- high growth, low share
- low growth, low share

This simple categorization system can help you to categorize a variety of data and values. You can then determine which to drop or which to take immediate action on, for instance, based on which quadrant they fall into. In keeping with that, the above example shows a 3x3 matrix that can be used in a sequence.

## *Ranges, Comparisons, and Evaluation*

The conceptual framework can be used in many areas, as it operates on a sliding scale.

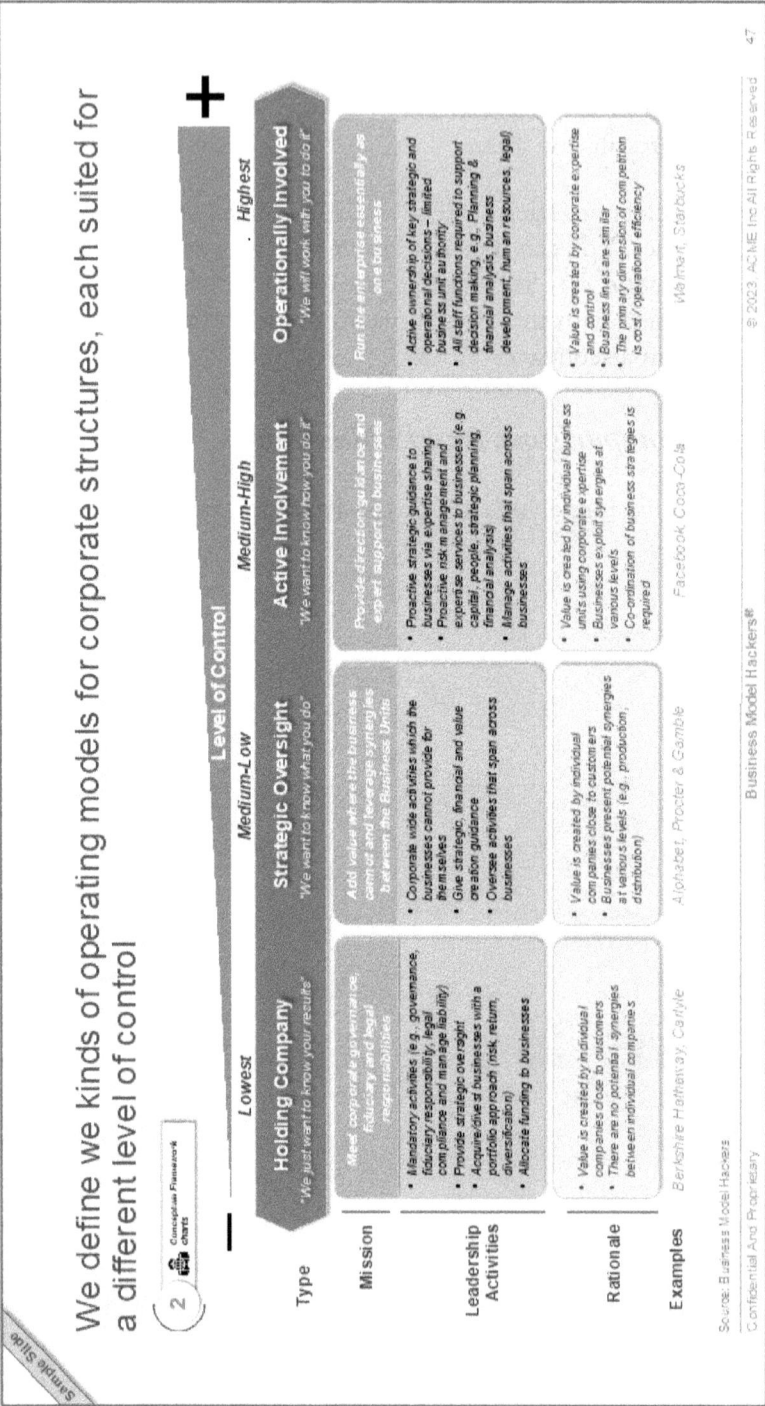

## We define we kinds of operating models for corporate structures, each suited for a different level of control

Conceptual Framework charts

**Level of Control**

| Type | Lowest — Holding Company "We just want to know your results" | Medium-Low — Strategic Oversight "We want to know what you do" | Medium-High — Active Involvement "We want to know how you do it" | Highest — Operationally involved "We will work with you to do it" |
|---|---|---|---|---|
| **Mission** | Meet corporate governance, fiduciary and legal responsibilities | Add value where the business cannot and leverage synergies between the Business Units | Provide direction, guidance and expert support to businesses | Run the enterprise essentially as one business |
| **Leadership Activities** | • Mandatory activities (e.g., governance, fiduciary responsibility, legal compliance and manage liability)<br>• Provide strategic oversight<br>• Acquire/divest businesses with a portfolio approach (risk, return, diversification)<br>• Allocate funding to businesses | • Corporate wide activities which the businesses cannot provide for themselves<br>• Give strategic, financial and value creation guidance<br>• Oversee activities that span across businesses | • Proactive strategic guidance to businesses via expertise sharing<br>• Proactive risk management and expertise services to businesses (e.g. capital, people, strategic planning, financial analysis)<br>• Manage activities that span across businesses | • Active ownership of key strategic and operational decisions – limited business unit authority<br>• All staff functions required to support decision making, e.g. Planning & financial analysis, business development, human resources, legal |
| **Rationale** | • Value is created by individual companies close to customers<br>• There are no potential synergies between individual companies | • Value is created by individual companies close to customers<br>• Businesses present potential synergies at various levels (e.g. production, distribution) | • Value is created by individual business units using corporate expertise<br>• Businesses exploit synergies at various levels<br>• Co-ordination of business strategies is required | • Value is created by corporate expertise and control<br>• Business lines are similar<br>• The primary dimension of competition is cost / operational efficiency |
| **Examples** | Berkshire Hathaway, Carlyle | Alphabet, Procter & Gamble | Facebook, Coca-Cola | Walmart, Starbucks |

Source: Business Model Hackers

For example, it can be used as a guide to select a model—in the above case, an operating model. To do this, the framework will have you define your scale, as shown above, and then use that scale to help you choose the adequate operating model for you, as shown below:

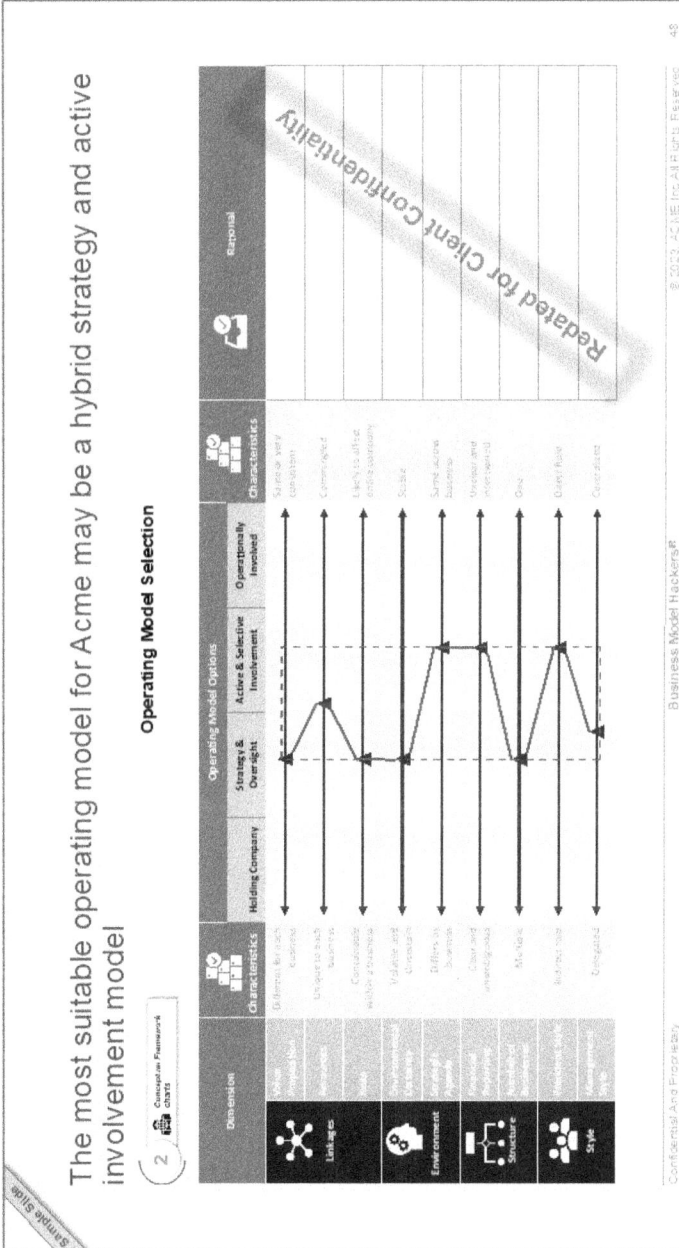

Alternatively, you can use the scale you have defined as a gap assessment tool, especially if you're evaluating how something is performing against the "best practice" in the industry or something of the like. The fact that these charts veer toward simplicity means that they can be designed as visuals able to make powerful analyses and deliver thorough insights.

# A number of challenges and issues were identified across 6 of the 8 key performance factors

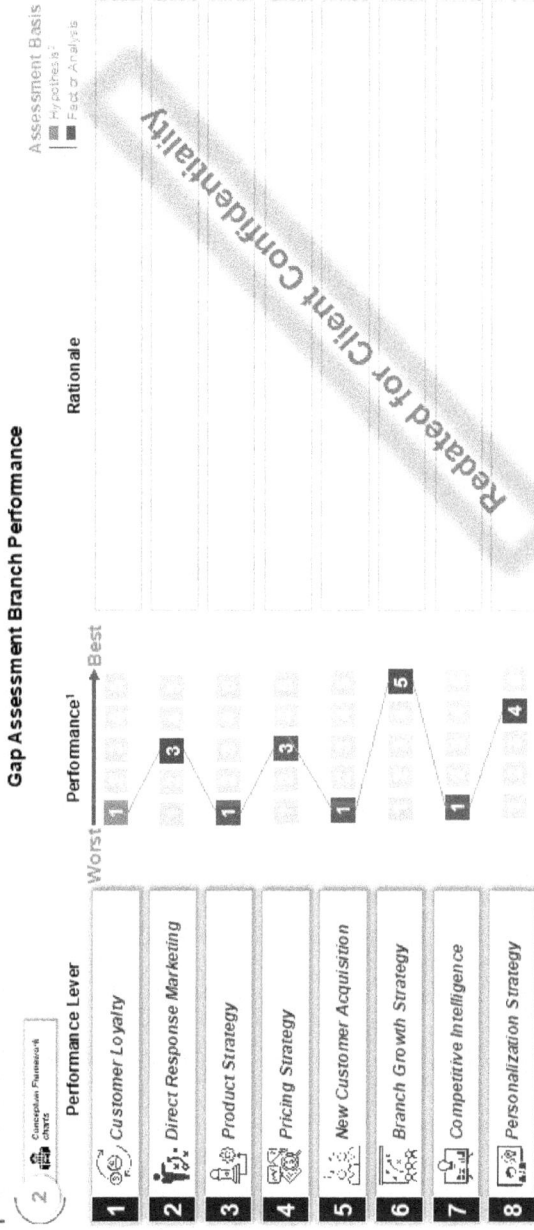

**Gap Assessment Branch Performance**

Performance Lever — Performance[1] (Worst → Best) — Rationale

Assessment Basis: Hypothesis[2] · Fact or Analysis

1. Customer Loyalty
2. Direct Response Marketing
3. Product Strategy
4. Pricing Strategy
5. New Customer Acquisition
6. Branch Growth Strategy
7. Competitive Intelligence
8. Personalization Strategy

Redacted for Client Confidentiality

Sample slide · Conception Framework charts

1 Performance is relative against "best in class"; 2 Hypotheses are preliminary and are awaiting validation with quantitative survey. Source Business Model Hackers; Team analysis

## *The Story Flow Framework*

The structure visual is one of the most important visuals that can be used in any story. As per its name, it's used to structure the flow of either a whole story or of stories within the larger narrative. It's a better alternative to turn to than bloaters and separators, seeing as it helps maintain the horizontal flow without introducing any interruptions to it.

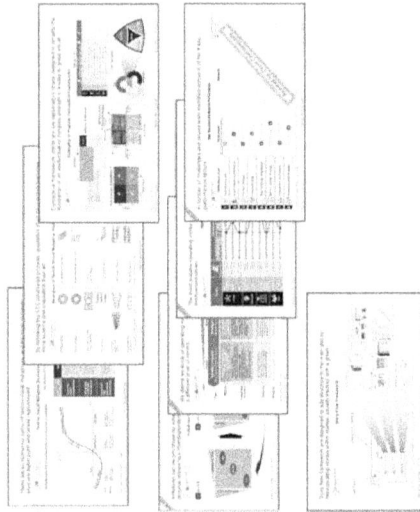

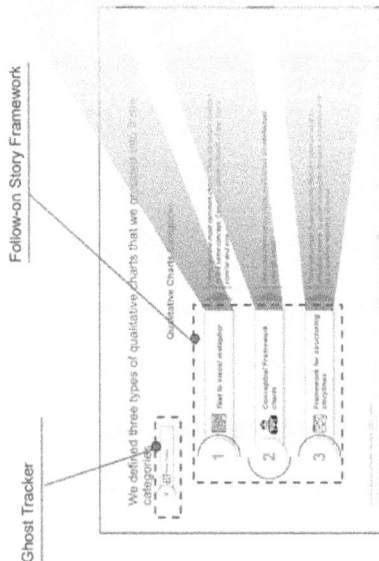

## *Qual Combo Charts*

Qual Combo Charts include both qualitative and quantitative visualizations on a single slide. They are particularly useful in highlighting, expanding, and explaining the takeaway the audience is supposed to make when looking at a quantitative chart. They come in especially handy when the takeaway can't be intuitively interpreted by the audience. They also make sure that the audience is able to directly access the claim that is being made or the message that is being given.

# Recap

- The Lindy Effect dictates that certain things really do age in reverse. There are two different kinds of things that do this: perishable and nonperishable things.
- STC is nonperishable and in fact shows signs of aging in reverse.
- You start STC with your horizontal logic first and then move on to your vertical logic, which determines your chart's blueprint.
- Your horizontal logic defines the kind of analysis and data you need to back up a claim or a message.
- Always work your way from your hypothesis or claim and have the visual in mind before you jump to data mining.
- Always avoid working backward in STC.
- The right strategy to adopt when building a story with charts is to begin by figuring out your goal, strategic objectives, and big idea.
- The most important skill you'll ever need in STC is the ability to ask the right questions.
- Always be consistent throughout your presentation. Having a template to work from can ensure that all the chart rules remain uniform across your deck.
- If you hide the headline and only view the chart, you should be able to guess what the headline says from the content in the chart and vice versa.
- The fewer pages in your templates, the better and the more consistent your deck's "look and feel" will be.
- The best stories flow seamlessly from beginning to end without being interrupted by bloaters. If your story doesn't flow well, you need to revisit your storyline.
- 5SUF starts with the hypothesis. From there you develop your key question.
- Next, you answer that question in your headline, which helps you to develop your blueprint.

- You can now use your blueprint as a guide for your analysis, which will in turn narrow down your data requirements.
- Finally, the narrowed down and chosen data will be displayed in a compelling visual in the form of a chart.
- You can make use of both quantitative charts and qualitative charts. As a rule, unless you're working in the legal field, you should try to use quantitative charts as numbers speak louder than words.
- When applicable, use Fermi Thinking to check if your data is within a reasonable range.
- Sensitivity analysis is an analysis that's made by analyzing the movement of a specific metric under different values. By contrast, scenario analysis is an analysis that's made by analyzing the movement of a specific metric in different scenarios. You should run both types of analysis if you have a variance at hand, so that you can see whether your claims hold up or not.
- The fewer words that make up your claims, the better the charts.
- The TVMA blueprint may lead you to multiple chart blueprint options. When in doubt, plot multiple versions of the data and pick the visual that's most suited to it.
- Visual cues help guide the reader's focus to the part of a chart that reinforces the message its headline is trying to give and can serve as memory cues.

# Chapter 3: Tips, Tricks and Best Practices

*"Design is not just what it looks like and feels like. Design is how it works." -*
*Steve Jobs*

Do you ever wonder why some slide designs seem effective while others appear more confusing — even messy? It often comes down to principles that were identified more than a century ago, on account of an inquisitive psychologist on a train ride.

In 1910, Max Wertheimer was traveling from Vienna, ostensibly on holiday. But at the train station, his attention was caught by an observation: the manner in which strobe lights, flashing in succession, created an illusion of movement. It was not only a trick of the light; it was a trick of the brain. He later named it the phi phenomenon (Schultz & Schultz, 2015).

Curious, Wertheimer probed further, using an instrument called a tachistoscope to precisely time and control visual stimuli. His experiments showed something radical for the time: our perception is not simply the sum of sensory inputs. The brain actively organizes what it sees (Wertheimer, 1912). This insight contradicted the dominant 'structuralist' theories.

When Wertheimer returned to Frankfurt, he joined forces with Wolfgang Köhler and Kurt Koffka. Together, they established the principles that underlie Gestalt psychology — the idea that we instinctively perceive objects in organized wholes, patterns, and recognizable forms. Their famous maxim that "the whole is greater than the sum of its parts," was not a mere academic musing (Koffka, 1935). It has deep implications for anyone trying to communicate visually — with charts and especially, with data.

## Intuition Codified: The Gestalt Principles

Gestalt psychology provides an interesting glimpse into how our minds instinctively make sense of the visual world by grouping and interpreting objects and elements without us consciously thinking about them (Köhler, 1947). Although academics have listed as many as a dozen specific principles (Schultz & Schultz, 2015), we won't delve into all the theoretical particulars here. When designing for clarity and impact in charts, I find it helpful to consider these as powerful guidelines rooted in how people naturally see.

Don't take every nuance as the Holy Grail. Instead, I'd recommend internalizing the following five principles as you design charts. It'll help you create slides that feel intuitive, direct the viewer's focus, and communicate your message more effectively:

4. **Proximity**: Things that are close to each other look like they belong together. It's that simple. Our eyes naturally group nearby items.

    ⇒ Use this by: Putting your chart title immediately above the chart, positioning labels close to the data points they describe, or physically grouping similar metrics on the slide. Avoid making the viewer's eye jump across blank space to connect related ideas.

5. **Similarity**: We also place things that look alike together. A relationship is signaled by shared color, shape, size, or orientation.

    ⇒ Use this when representing the same data series in multiple charts by using a consistent color across all charts for the that data series. Similar elements should ideally have the same shape or font style. This consistency creates visual coherence and facilitates comparisons.

6. **Closure:** Our brains are accustomed to completeness. In other words, we often fill in gaps to make them relate to more recognizable shapes.

    ⇒ Use this by: Simplifying the visual but making it recognizable. Often, an outline or silhouette communicates a more complicated concept (for example, a customer segment) better than a detailed illustration can, allowing the audience's mind to 'fill in' the shape.

7. **Continuation**: The eye wants to follow lines, curves, and paths. Once the eye establishes a visual direction, it tends to continue down that same trajectory.

⇒ Use this by: Placing elements on a well-defined visual path (such as the natural Z-pattern or F-pattern of reading) to help draw attention. Use smooth lines in a line chart, so the eye intuitively follows the trend.

8. **Figure-Ground**: We automatically separate a main object (the figure) from its background. Clear distinction is key.

⇒ Use this by: Providing a strong contrast between essential elements (such as text or data points) and the background. Light text over a dark background, or the other way around, improves reading ease and makes the important 'figure' stand out.

By remembering these Gestalt principles, you can focus on more than just placing data on a slide. You can begin designing visuals that not only amplify your vertical logic, but also anticipate how your audience will interpret aesthetics, making your data stories even more convincing.

# Chart and Slide Design Best Practices: The Nuts and Bolts

**Putting Principles into Practice**: Step one is to understand how people see (The Gestalt Principles). Step two is putting that knowledge into action with tangible design practices. If you've downloaded the Storytelling Charts Add-In (STCAI) or templates from our toolkit, most of the basic formatting has already been done for you. Those tools are based on best practices. But if you understand why these practices work, and how they should and shouldn't be used, it prepares you to communicate more effectively.

Here are some of the most important principles to bear in mind at all times:

- **Unit**: Don't make it hard for your audience. Never put too many digits in a chart. As a general rule of thumb, no more than 3 digits should be displayed. For millions (10 million to 99 million), use "96.5

(in millions)". Once you cross triple-digit millions, it's the billions (e.g. 129,900,000 -> "0.13 (in billions)). And when rounding, prioritize accuracy together with digestibility.

- **Consistency**: Think of consistency as the visual 'voice' of your slide deck. Be consistent with color palettes, font choices, and chart types for similar data throughout your entire deck or report. A cohesive look and feel enables your audience to concentrate on the insights rather than getting lost in jarring stylistic discontinuities.

- **Simplicity**: Don't overcomplicate things. As Steve Jobs hinted, "good design works". Unneeded 3D effects, and distracting gradients — contribute to unclear visuals. Just remember to take the "less is more" approach. A clean chart without clutter is almost always a more effective and professional.

- **Clarity**: If your audience is squinting, guessing, or hunting for a meaning, your chart isn't clear enough. Labeling needs to be clear and concise. It should be clear what axes, data series, legends, units, and dates mean. Use clear titles and subtitles to establish context early. Make use of annotations or use labels to show critical findings right on the chart.

- **Emphasis**: Don't force your audience to work hard to see your main point. Use subtle ways to lead your audience where to look. Complementary colors (in moderation!), bold fonts, arrows, or callouts can direct attention to the most important data. The effective use of whitespace and intentional placement creates a visual hierarchy that guides the viewer through the data in a logical manner.

- **Accessibility**: Good design is inclusive design. Some people have impairments in their vision, so make sure your charts are legible to them as well. Use enough color contrast (test this!). Use long and descriptive alternative text for screen readers. Another important point to check is how your charts look when printed out in black and white – a common scenario! Another common design trap is relying on color alone for differentiation. Making your charts accessible to all isn't simply good practice; it helps ensure your message reaches the widest possible audience.

# The Ultimate Tests: "So What?" and the Clock

Beyond those mechanics of visualization, there are two final checks for every single chart you create:

## *The "So What?" Factor*

This is the ultimate test. After you build your chart, ask yourself: **So what?**

How does the insight presented on this slide eloquently lead and/or relate back to the big-picture and overarching strategic goals and objectives of your presentation or deck? If the answer is weak, tangential, or nonexistent, that chart should either be rewritten or eliminated. In other words, if it doesn't add to your core message in a meaningful way, it's noise, no matter how good it looks. Each chart must justify itself by helping to answer that all-important "So what?" question for your audience.

## *The 3-5-10-15 Rule*

Value your audience's time and mental bandwidth. They should be able to understand the main takeaway of a slide fairly quickly (Cowan, 2010). As a guideline:
- 3 seconds for slides that are very simple (e.g., one important number)
- 5 seconds for somewhat more complex visuals (e.g., simple bar chart)
- 10 seconds for medium-complexity slides (such as a complex line chart with a few series).
- No more than 15 seconds for the most complex visuals you include.
- Test this! So, before you finalize: test slides on a colleague who is not familiar with the content. If they have difficulty understanding the main point within these time limits, the slide is probably too complicated. Make it shorter, spread it out over multiple slides, or add more obvious emphasis.

# Recap
- The Gestalt Principles provide practical guidelines for effective visual design: proximity, similarity, closure, continuation, and figure-ground
- Focus on these principles as rules of thumb rather than strict requirements
- When displaying large numbers, avoid showing more than 3 digits on charts
- Maintain consistency in style across all charts in your presentation
- Adopt the "less is more" principle in chart design
- Ensure clear and concise labeling of all chart elements

- Use strategic techniques to emphasize key takeaways
- Create accessible charts for all audiences, including those with visual impairments
- Apply the "So What?" test to every slide to verify it connects to your objectives
- Use the 3-5-10-15 Rule: audience should grasp simple slides in 3 seconds, slightly complex in 5 seconds, moderately complex in 10 seconds, and most complex in 15 seconds
- If viewers struggle to understand within these timeframes, simplify your slide or break it into multiple slides
- Test your slides with someone before finalizing your presentation

# Chapter 4: Quantitative Chart Examples

In this final chapter, we'll focus more on practical applications. I thought, what better way to close this book than by adding a section on the most frequently used quantitative chart examples. The idea here is to be able to create charts effortlessly and intuitively when you know you have a specific message and its corresponding data to back it up.

In the previous chapters, I provided a lookup table or cheat sheet that guides the type of chart to use by context. Here, I will provide a reference by chart type for the most frequently used charts and the situations they are typically used in.

Throughout this book, we've explored the fundamentals of Storytelling Charts (STC), delving into the psychological aspects of vertical logic, horizontal logic, and best practices for creating compelling and persuasive slide presentations. We've discussed the importance of understanding your audience, crafting a clear narrative, and using the right chart types to convey your message effectively.

Now, it's time to put all that knowledge into practice. In this chapter, we'll walk through a series of real-world examples, showcasing how to apply the principles of STC to create impactful and memorable presentations. We'll explore a variety of chart types, including flag charts, line charts, and more, discussing when and how to use each one for maximum effectiveness.

By the end of this chapter, you'll have a comprehensive toolkit of practical examples and chart types to draw from as you create your own presentations. Whether you're presenting to colleagues, clients, or stakeholders, you'll be equipped with the skills and knowledge needed to craft compelling stories that inform, engage, and persuade your audience.

Now, let's dive in and see how the principles of Storytelling Charts and vertical logic come to life in real-world applications.

# Flag Charts or Horizontal Bar Charts

Now let's talk about a mighty workhorse of the charting world; the horizontal bar chart, also known as the "Flag Chart". Why "flag"? Because the bars often protrude from the y-axis and resemble tiny flags on poles. But the name doesn't matter as much as the action. These types of charts are great any time you want to compare items to each other — like performance across regions, survey responses across a category, or benchmarks against competitors.

Their superpower? That horizontal layout. It provides lots of room for readable labels next to each of the bars, which quickly becomes difficult in vertical column charts as soon as you have more than a handful of categories and longer names.

Flag Charts can be as simple as bars next to each other or stacked (segmented), so they are versatile.

Here are some of the common scenarios where flag charts perform gracefully:

## *Showing Two Sides: The Diverging Bar Chart or Tornado Chart*

At times, you want to present not only a number, but the extent to which it strays from a middle point, or contrast two competing categories with one another directly. This is why the diverging flag chart is useful.

- **Example: That Customer Satisfaction Survey** — Imagine asking your customers to rate various aspects of your service on a nonlinear scale from −5 (highly dissatisfied) to +5 (highly satisfied). Here we have an ideal case for a diverging chart. You would set your baseline at '0' (neutral). Positive ratings (such as for Product Quality) extend as bars to the right, while negative ratings (perhaps for Customer Service) stretch to the left. In an instant, your audience knows where you're succeeding and where you need work. No need to sift through figures.

- **Example: Population Structure (The "Tornado"):** You've probably seen pyramid graphs that are used to compare the number of males and females in each age group. That's often a tornado chart — a kind of diverging chart with back-to-back sets of flag charts that typically share the same y-axis (age groups). It takes its name from that distinctive funnel or butterfly wing shape, which creates a striking visual summary of population distribution.

## *Visualizing Spreads and Timeline: Range Visualization*

**Customer Satisfaction Survey Results**
(Number of Respondents = 500)

| Highly Dissatisfied | -5 | -4 | -3 | -2 | -1 | +1 | +2 | +3 | +4 | +5 | Highly Satisfied |

| Product Quality | 4 |
| Pricing | 3 |
| Website Usability | -1 |
| Customer Service | -2 |

Flag charts can also be used to show ranges, particularly time, and are not limited to single values.

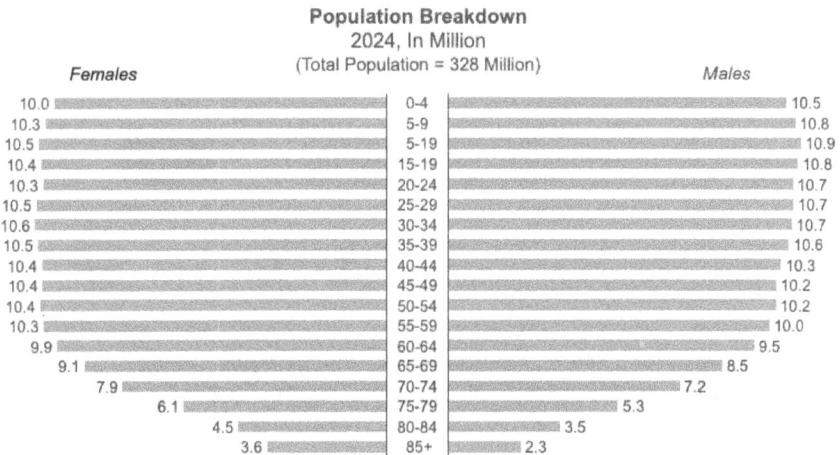

**Population Breakdown**
2024, In Million
(Total Population = 328 Million)

| Females | | Males |
|---|---|---|
| 10.0 | 0-4 | 10.5 |
| 10.3 | 5-9 | 10.8 |
| 10.5 | 5-19 | 10.9 |
| 10.4 | 15-19 | 10.8 |
| 10.3 | 20-24 | 10.7 |
| 10.5 | 25-29 | 10.7 |
| 10.6 | 30-34 | 10.7 |
| 10.5 | 35-39 | 10.6 |
| 10.4 | 40-44 | 10.3 |
| 10.4 | 45-49 | 10.2 |
| 10.4 | 50-54 | 10.2 |
| 10.3 | 55-59 | 10.0 |
| 9.9 | 60-64 | 9.5 |
| 9.1 | 65-69 | 8.5 |
| 7.9 | 70-74 | 7.2 |
| 6.1 | 75-79 | 5.3 |
| 4.5 | 80-84 | 3.5 |
| 3.6 | 85+ | 2.3 |

- **Example: Keeping Projects on Track (Gantt charts):** If you manage projects, you are familiar with the Gantt chart. It's basically a collection of flag charts! Tasks are horizontal bars. The left edge indicates the start date, the right edge indicates the end date, and the length of the bar indicates the duration. It helps you see at-a-glance what's coming up with your project — what overlaps, what's dependent, what's next. It's the heartbeat of the project in visual form.

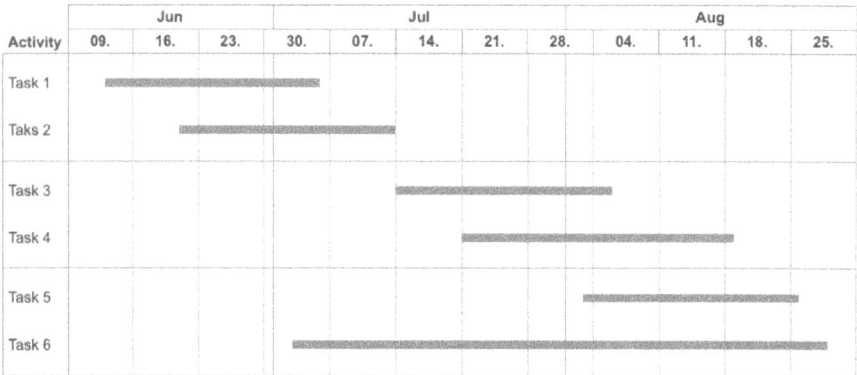

| Activity | Jun 09. | 16. | 23. | 30. | Jul 07. | 14. | 21. | 28. | Aug 04. | 11. | 18. | 25. |
|---|---|---|---|---|---|---|---|---|---|---|---|---|
| Task 1 | ▬ | ▬ | ▬ | | | | | | | | | |
| Taks 2 | | | ▬ | ▬ | ▬ | | | | | | | |
| Task 3 | | | | | | ▬ | ▬ | ▬ | | | | |
| Task 4 | | | | | | | | ▬ | ▬ | ▬ | | |
| Task 5 | | | | | | | | | | ▬ | ▬ | |
| Task 6 | | | | | ▬ | ▬ | ▬ | ▬ | ▬ | ▬ | ▬ | ▬ |

## *Spotting Winners and Losers: Deviation and Ranking*

Flag charts can be used for visualizing benchmarks, e.g., the performance of different regions, products, or teams against each other. Or just ranking them from worst to best. This can be illustrated nicely with flag charts.

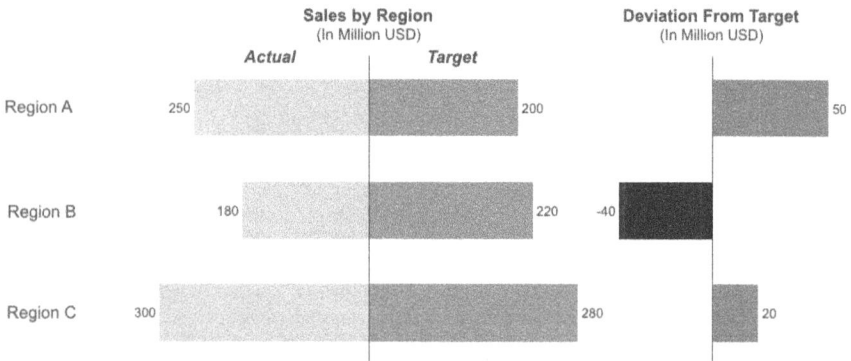

**Sales by Region** (In Million USD) — *Actual* / *Target*; **Deviation From Target** (In Million USD)

| Region | Actual | Target | Deviation |
|---|---|---|---|
| Region A | 250 | 200 | 50 |
| Region B | 180 | 220 | -40 |
| Region C | 300 | 280 | 20 |

- **Example: Sales Performance vs. Target:** Imagine you are comparing the sales from different regions against a target. A flag chart can be creatively used to visualize this comparison. In this example, the flag chart would display the sales figures as bars extending from the target baseline. Regions that exceeded the target (Regions A and C) would have bars extending to the right, while regions that fell short (Region B) would have bars extending to the left. By color-coding the bars and ordering the regions based on their deviation from the target, you could quickly identify top-performing and underperforming regions.

### *One Tip to Make the Most Impact*

One best practice that makes these comparison charts far easier to read is to order your data in descending or ascending order (assuming, of course, there is no natural order — like time). It helps the viewers' eyes to travel fast and makes the relative differences easier to process without extra effort.

Flag charts are very useful in their multitude of forms, as they leverage the human's natural ability to compare lengths and positions horizontally. By knowing how to use these styles — from flagging deviations, to timelines, to ranking performance — you have a powerful and flexible solution in your arsenal, one that will ensure your data stories are clear, compelling, and instantly grasped by your audience.

# Column Charts: Standing Up for Comparison

Vertical bar charts, also known as column charts, are versatile tools for visualizing data across various categories or time periods. They can be used to show time series data, compare values across categories, or display the relationship between different data series over time. Column charts display data using vertical bars, with the height of each bar representing the value for a specific category or time period.

# Time Series Column Charts

Time series column charts are ideal for showing changes or trends over time, especially when you have a relatively small number of data points (usually less than 20). They help visualize data with a clear beginning and end for each time period, such as monthly sales or quarterly revenue.

- **Example: World Poverty Ratio:** To show how poverty has evolved over the past 40 years, a line chart could be used. However, a time series column chart would also be a great choice. In this example, each year would be represented by a vertical column, with the height of the column indicating the percentage of people living on under $2.15 per day at 2017 prices. By isolating specific periods and adding Compounding Annual Growth Rate (CAGR) to these periods, you can clearly visualize how the trend shows a slow drop (-1.6%) between 1981 and 1990. The trend shows a rapid acceleration (-5.1%) in poverty reduction after 1990, up until 2018. Since 2018, a complete reversal of the trend occurs, and we start to see an increase in poverty rates by 0.6% since.

## Grouped Column Charts

Grouped column charts are used to show the relationship between different data series over time or across categories. Each group of columns represents a specific time period or category, and within each group, there are multiple columns representing different data series.

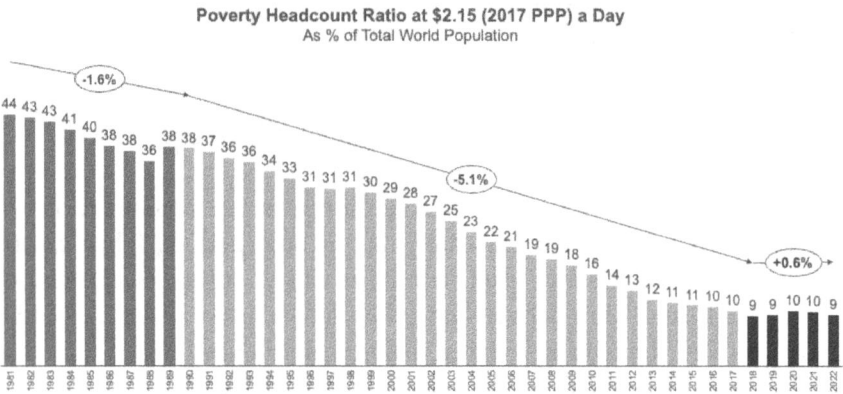

Poverty Headcount Ratio at $2.15 (2017 PPP) a Day
As % of Total World Population

- **Example: Quarterly Sales by Region:** Imagine you want to compare the quarterly sales performance of different regions over the past four years as shown below. In this example, each region would be represented by a group of four columns (one of each year), with each column within the group representing the sales value for a specific region over the years. This allows you to compare the sales growth performance for each of the regions and observe trends or changes over time.

In this visualization the key takeaway is that Region B has shown consistent growth trend.

## *Stacked Column Charts*

Stacked column charts are used to display the individual components that contribute to a total value. Each column represents the total value, and it is divided into segments representing the different components.

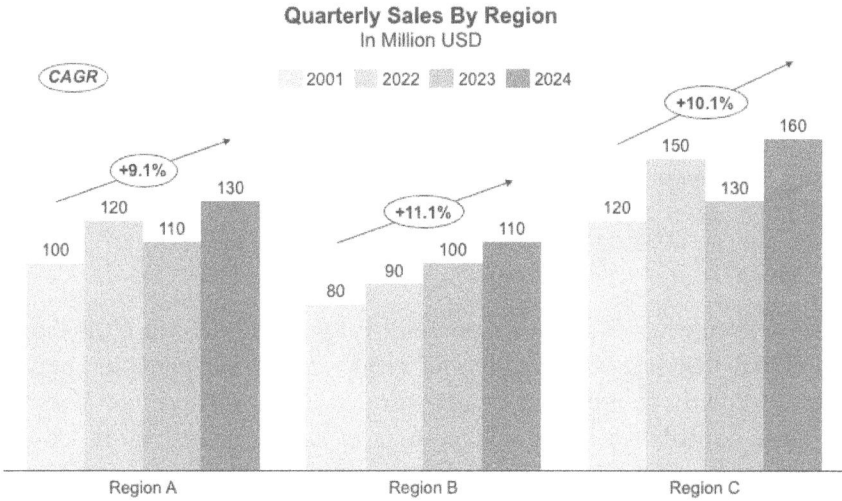

**Quarterly Sales By Region**
In Million USD

These charts help reveal more insights, such as breakdown in each stacked column as well as overall trends by component and total. To illustrate, let's take the same example from the Grouped column charts and visualize the data in stacked column charts.

- **Example: Quarterly Sales by Region:** As you can see in the chart below, each year would be represented by a vertical column, with the height of the column indicating the total sales for that year. The column would be divided into three segments, each representing the sales contribution of a specific region (Regions A, B, and C). This allows you to see the total sales for each year and understand how each region contributes to the overall sales. The visual reveals several components, including sales per region per year. This would be useful to visualize overall CAGR vs. regional CAGR, as well as the total sales per year. These were not identifiable in the Grouped column visual.

**Quarterly Sales By Region**
In Million USD

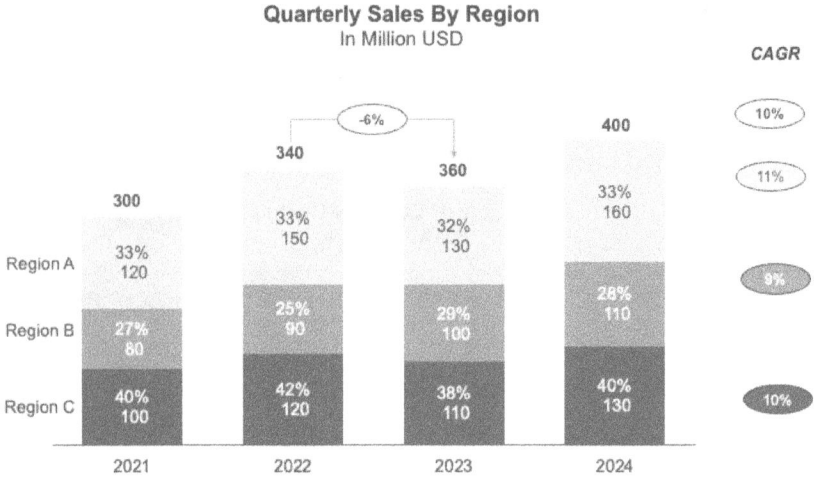

The key takeaways in this version of the column chart are some additional insights not seen in the grouped column version, such as overall growth, the drop in overall sales in 2023, and the fact that the regional contribution to the overall sales has remained more or less constant.

One best practice to note about Stacked Column Charts is always display the segments in descending order (i.e., from the bottom to the top with the largest segment at the bottom of the column).

## *Comparison Column Charts*

In Column Charts, the height of bars illustrates relationships between categories. They're commonly used to compare discrete groups on the same measure, like salaries of different CEOs. They are universally understood and excellent for simple category comparisons. However, many bars may create the impression of a trend line rather than highlight discrete values, and multiple groups of bars can become difficult to interpret. To address this, the columns could be butted against one another to make it easier to compare the relative heights of the bars.

In a way, they are similar to flag charts. Both Flag and Column Charts are essentially bar charts designed to facilitate visual comparisons between different data points or categories. The horizontal orientation of these flag charts allows for the inclusion of less data points, albeit with longer labels. On the other hand, column charts are particularly useful when dealing with shorter labels and more data points. Additionally, the vertical layout of column charts provides space for additional content such as callouts, percentage indicators, or visual elements that can enhance the chart's informational value.

In the following example of General Government Debt to GDP by Nation, you can isolate outliers like Japan and Germany or highlight specific countries.

**Debt to GDP**
In %, 2022

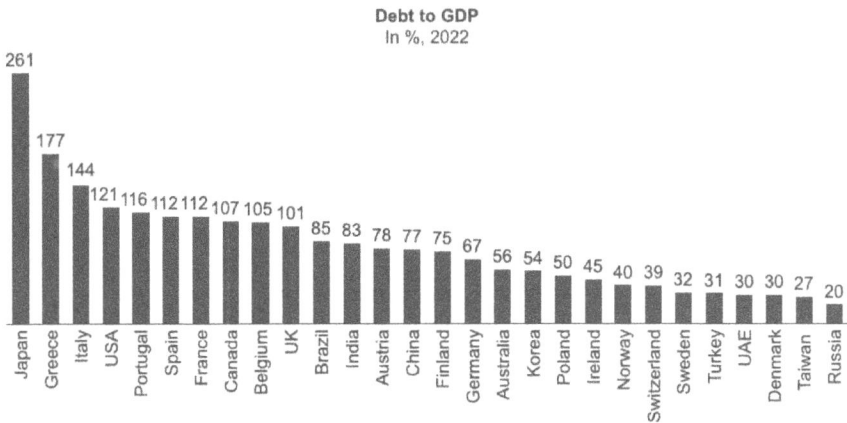

These examples illustrate the effectiveness of column charts for visualizing data across categories, time periods, and different data series. By understanding the various types of column charts and when to use them, you can create informative and visually appealing charts that effectively communicate your data insights to your audience.

One best practice to note about Comparison Column Charts that are not time dependent is to always display the data in descending order.

# Line charts: Telling the Story of Change Over Time

If there is one kind of chart that almost everyone knows, it's this one. Consider the basic stock market line graph depicting its highs and lows, or a chart plotting temperature changes over the course of the year. We sometimes refer to them as "fever charts" or "trend lines," and that gets directly to one of their key strengths: displaying how something evolves over a consecutive span. The magic is in connecting the dots. By connecting individual data points (such as daily stock prices or monthly sales figures) with a line, they create a flowing visual narrative. Your eye just automatically follows it, which makes trends and patterns easy to spot intuitively so you can get a feel for the data's overall journey. Is it generally going up? Down? Staying flat? Was there a strange event at a particular moment? These larger insights are immediately clear from the line chart.

However, while they highlight the general trend, sometimes the precise value of any given point along the graph gets a little lost in the visual sweep of the trend. And there's a notorious trap: trying to put too many lines on one chart run the risk of the "spaghetti chart" — a spaghetti-like confusion where it is impossible to follow anything clearly. However, if used wisely, the line chart is a foundation of data storytelling. Let's look at how.

## *The Classic: Time Series Line Charts*

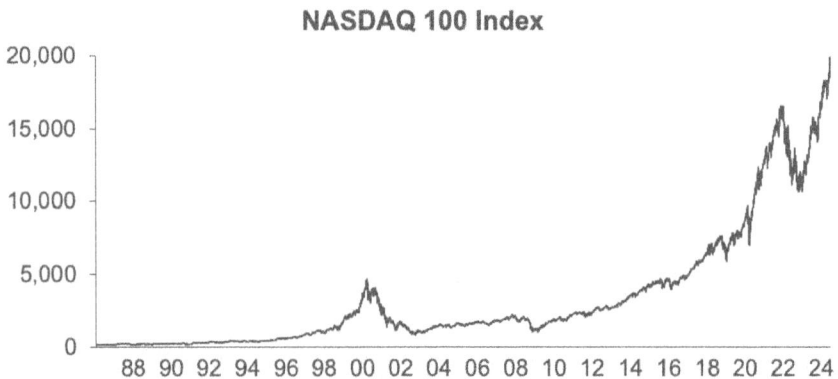

This is the bread-and-butter use case. The time series line chart is your best option when you need to demonstrate how something has progressed over days, months, years, or even decades, particularly when you have thousands of data points. It makes the noise vanish and enhances the continuity, the story arc, of your data.

- **For example, the NASDAQ**: Imagine plotting the history of the daily closing value of the NASDAQ index going back to the 1980s.

Plot each day as a point and connect them to show the dramatic growth spurts, the bubbles, the crashes, and the recoveries over decades. It's not about being precisely correct about the value and the movement of June 5th, 1992 — you're watching the whole economic story unfold.

## *Hold On – Let's Talk Aspect Ratio (it's more important than you think!)*

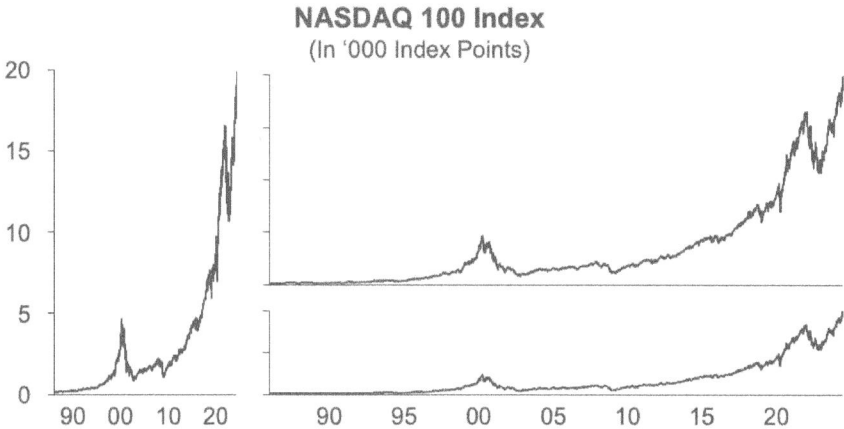

**NASDAQ 100 Index**
(In '000 Index Points)

There's one special, critical thing about line charts: the aspect ratio, that is, the ratio between the width of the chart and its height. This isn't only an aesthetic issue — it completely shifts the way your audience views the trends.

- **Wide and Flat**: Shows a gradual, less dramatic approach to change.
- **Tall and Skinny**: Exaggerates changes, making slopes steeper, and trends appear more erratic.

Remember that NASDAQ chart? If we squeeze it horizontally (make it tall), the market spikes are terrifyingly pronounced. Stretch it out horizontally, and those same swings appear much tamer. So, what's the "right" ratio?

There's no universal magic number, but there are two key principles:

- **Be Honest:** Don't mess with the aspect ratio just to make a trend look more dramatic or less alarming than it actually is. That's misleading.
- **Be Consistent (and Conventional for Time):** Use similar overall aspect ratios for your related charts, one that is typically consistent.

For time series data, the default convention is wider than tall. This guidance is consistent with how we visualize time passing from left to right, and it makes charts easier to read and interpret for most audiences.

## Comparing Journeys: Comparison Line Charts

**Real Returns on Major Asset Classes (In %)**
Q1-2020 = Baseline

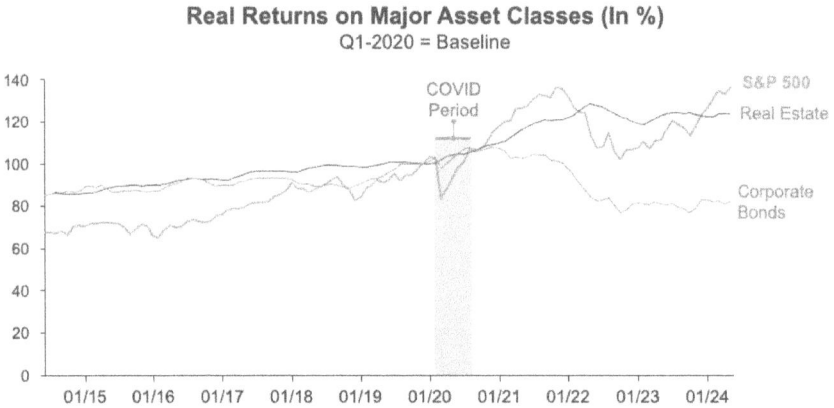

So what if you want to see the trends for different items next to each other? This is where comparison line charts come into play. Plotting multiple lines on the same axes allows you to see how the different categories or data series compare against each other.

- **Example: Investment Returns:** Let's compare the real (inflation-adjusted) returns of investing in stocks (S&P 500), real estate, and corporate bonds over several years, including the pandemic. Plotting them as separate lines on the same chart tells the story: Stocks swooned first, then sprung back sharply, then cooled; real estate remained mostly steady but growing; and bonds lagged in the inflation-fed landscape. The lines soaring steep, then crossing and diverging, tell a comparative story that raw numbers would not capture as clearly as this visualization does.

## (A Quick Aside: Stacked Line Charts)

You may also encounter stacked line charts — line charts that represent components of a whole, layered above each other — similar to stacked columns. The space between the lines indicates how much each component contributes. They can work, but to be honest, they frequently get harder to read correctly than stacked bars or other chart types that show composition. Use them with caution.

## The Danger Zone: Overplotted or "Spaghetti" Charts

The allure of seeing just a few more lines at once quickly leads to disaster. A chart with eight trend lines is not twice as informative as one with four — it's often twice as confounding.

- Example: Women in Parliament: Consider attempting to display on one line chart how many percentage points of parliamentary seats in 15+ European countries were held by women over 20 years? It quickly becomes a jumbled flurry of crossing lines. You cannot follow individual countries, the colors become indistinguishable, and the overall message is entirely swallowed up in visual noise.

### How to Tame the Spaghetti

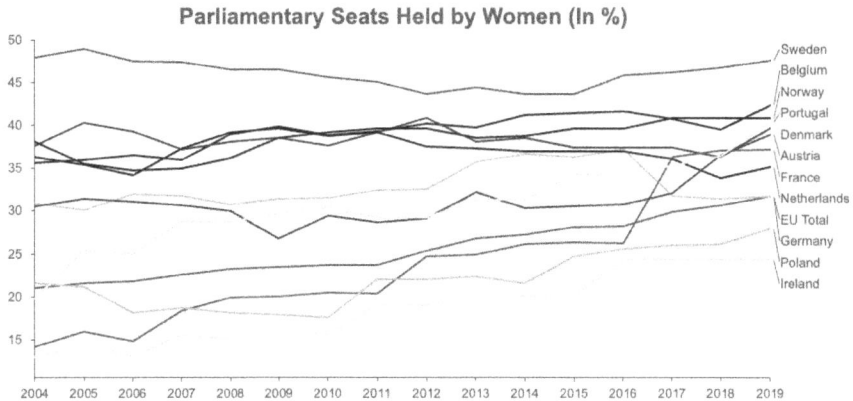

Parliamentary Seats Held by Women (In %)

If you've got a little too many lines, don't worry! You have options:

1. **Limit Lines**: Be ruthless. Can you describe the core story with 4–6 lines? That's generally the sweet spot for clarity on one chart.
2. **Use Contrast:** Use a heavier weight or brighter color to differentiate your main line(s). Make noncritical lines thinner and mute colors (say gray or dashed patterns). Guide the viewer's eye!
3. **Highlight & Annotate**: If you need to get all these lines on the page, consider graying out most of them and actively highlighting one or two trends that you want to discuss. Then use callouts or annotations to refer directly to meaningful events or takeaways written on the highlighted lines.

4. **Break It Down (Into Small Multiples):** This is usually the most graceful solution. Do not create one complex chart but create a grid of small, simple line charts (the so-called "Trellis Charts" or "Panel Charts"). Each of the small charts has the same axes and scale but includes only one or a few categories.

**Parliamentary Seats Held by Women (In %)**

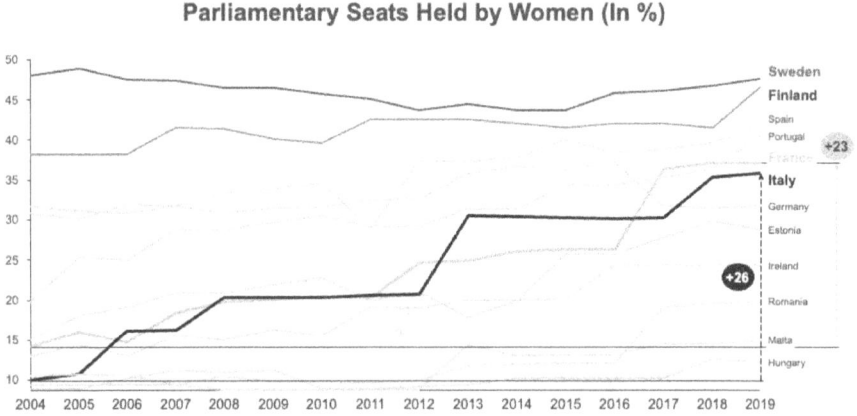

By highlighting the key trends, we emphasize in the example below the countries with the highest female participation in parliament (i.e., Finland and Sweden) and the ones with the highest increase (e.g., France and Italy).

As a ROT, keep in mind that the goal of a line chart is to clearly communicate trends and patterns. If the chart becomes so complex that it obscures this goal, it's time to reconsider your approach. Sometimes, creating multiple simpler charts is more effective than trying to cram everything into a single, complex visualization.

**Parliamentary Seats Held by Women By Country Vs. EU Total**
2004-2019, In %

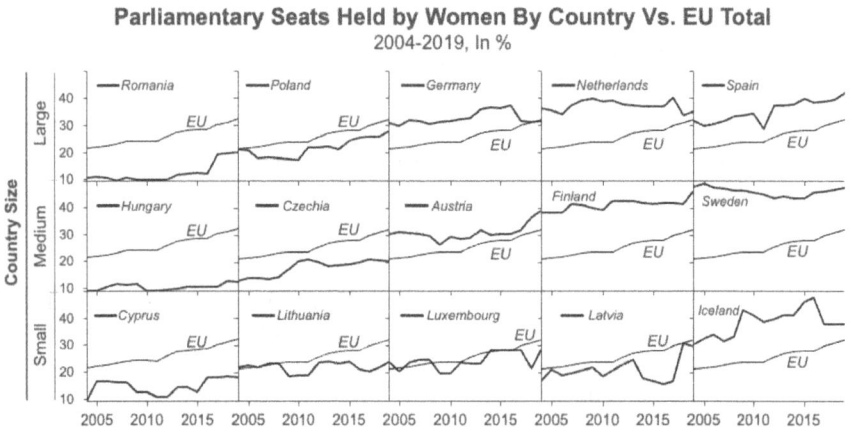

## Small multiples: Clarity in numbers

So back to that ugly Parliament data. Rather than one messy chart, picture a grid, with each little chart showing the trend in the EU average compared to the trend in one single country.

All of a sudden, you can see the trajectory of each country in comparison to the average and each other without the visual clutter. Small multiples do a great job of clearly showing individual trends across many categories. The trade-off? It becomes more difficult to identify exact crossover points between the categories that might otherwise be apparent in a single (less cluttered) chart.

The key features of Small Multiples Charts:
- Shows individual source trends and total consumption.
- Visualizes the shifting mix of components over time.
- Maintains visibility of individual trends within the total.

## Wrapping Up Line Charts

Line charts are most powerful storytellers when it comes to depicting change and trends over time. Their visual flow is easy to understand. But keep in mind the aspect ratio affects perception and beware the spaghetti monster! Be judicious in the use of comparison lines; when the variables to compare are numerous, accept the clarity provided through highlighting, annotating, or the neatness provided by small multiples. With these strategies in your toolkit, you will be able to use the line chart to tell accurate and effective stories of trends and comparisons.

# 100% Stacked Bar Charts

100% Stacked Bar Charts use multiple rectangles divided into sections to represent proportions of a whole. They're frequently deployed to illustrate simple breakdowns of totals. Many consider them superior to pie charts, as they effectively show dominant versus non-dominant shares and can handle more categories. They work both horizontally and vertically. However, including too many categories or grouping multiple stacked bars can make them challenging to use for discerning differences and changes.

One way to use stacked bar charts is to visualize proportions instead of pie charts as in the example below shows an 80-20 rule-like visualization.

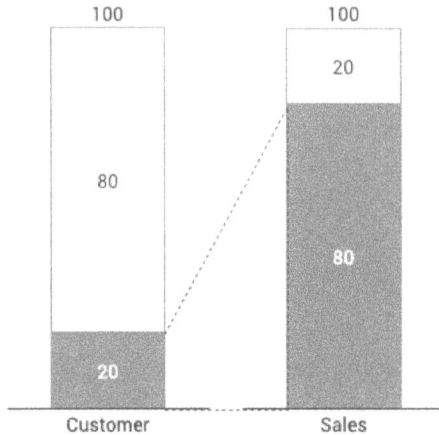

## Waterfall Charts: Explaining the Journey from Start to Finish

Ever have the need to show precisely how a beginning number led to an ending number? Or perhaps you need to demonstrate how an original sales figure became your actual profit after accounting for the various costs and adjustments in between. This is exactly where the Waterfall Charts shine.

You can think of it as building a bridge or following a cascade of water down a series of steps — hence the terms "bridge chart" or "cascade chart" that you sometimes hear. It visually guides your audience through a series of additions and subtractions. Unlike pie charts (discussed later) which provide a static snapshot of parts-of-a-whole, or stacked bar charts, which provide a view of composition, the waterfall is the king of showing the cumulative effect of an ordered series of changes. It's about the journey.

Waterfall charts have become a favorite among financial analysts for good reason, but their utility extends far beyond basic profit and loss statements.

## *How Does it Work Visually?*

Once you see one, it becomes quite intuitive:

- **The Starting Point**: You start with your base number – which could be last year's sales figures, starting inventory, or total revenue. This is typically the first full bar standing tall on the left.
- **The Steps (Ups and Downs):** Then follows the in-between numbers — the things that alter the original number. Positive contributions (e.g. new sales, new funding) are indicated by "floating" bars that rise up from the end point of the previous bar. Costs (negative contributions), expenses (negative contributions), and returns are shown as floating bars dropping down.
- **The Cumulative Flow**: This is the most important part – each floating bar starts where the previous one finished. You follow the running total visually, rising and falling as you traverse each step. It's like watching a tide come in and out.
- **The Grand Finale:** The full bar on the right represents the final result – the total, regardless of the number of steps it takes to reach the end.
- **(Optional) Milestones:** There are times when you want to display a subtotal along the way (think Gross Profit before Operating Expenses). For common breakpoints within the series, waterfall charts typically use a solid bar resting on the baseline to represent these key milestones or breakpoints in the sequence.

## *So Let Us Take That Profit & Loss Example:*

Imagine explaining Apple's profitability going from revenue to net income and how that might look in a waterfall chart using data extracted from Apple's Income Statement:

- **Revenue Streams**: Each tall blue bar represents one income stream (iPhone income is $201B, followed by Services $85B, Wearables & Home $40B, Mac $29B, and iPad $28B).
- **Total Revenue**: These positive impacts accumulate to reach Apple's Total Revenue ($383B), shown as a solid blue milestone bar.
- **Deductions**: From there, the downward red floating bars show deductions: Cost of Goods Sold (-$214B); Operating Expenses (-$55B); Taxes (-$19B).
- **Cumulative Effect:** Each red bar begins where the previous one ended, in sequence, deducting from the running total.
- **Net Income**: And this brings us to Net Income ($97B), represented by the final solid blue bar.

This waterfall chart elegantly visualizes how Apple's multiple revenue streams combine into total revenue, and how different costs and expenses then reduce this total to reach the final profit figure. It succinctly tells the financial tale of how Apple made almost $100 billion in profit, laying out revenue sources and cost categories all in one flowing visual.

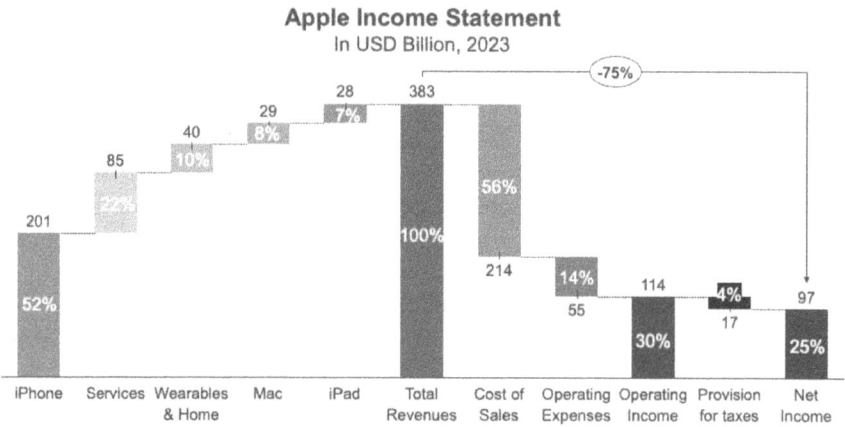

**Apple Income Statement**
In USD Billion, 2023

## When Should You Reach for a Waterfall Chart?

These charts are especially useful when you have to:

- Discuss two periods of performance differences between two periods (like year-over-year profit differences)
- Break down the components that contribute to a net change (e.g., explaining the variance between budget and actual results).
- Explain how various elements add up to the final value (see the P&L example).
- Illustrate trends in inventory, cash flow, or population over time.

The waterfall chart is, in many ways, the visual storyteller of financial and operational change. When you want to deconstruct the steps and demonstrate how positive and negative elements coalesce to achieve an ultimate outcome, it offers clarity and narrative flow that's difficult to beat.

# Scatter Plot Charts

Scatter Plot Charts, also known as dot charts or X-Y plots, display individual data points on a coordinate system and are used to visualize the relationship between two variables. Each dot represents a piece of data that has two numbers attached to it. That is, each data point is represented by a dot, with its position determined by its values on the horizontal (X) and vertical (Y) axes. For example, if you were looking at how height relates to weight, each dot might represent a person. The dot's position left-to-right would show their height, and its position up-and-down would show their weight. Scatter Plot Charts are particularly useful for identifying correlations, outliers, and patterns in data.

## *Correlation in Scatter Plots Charts*

Scatter plots are great for showing how two things or variables might be related and how strongly they are correlated.

Correlation is a statistical measure that reflects how two variables are related to each other. It doesn't necessarily imply cause and effect, but rather the extent to which they change together.

- **Positive Correlation**: If the dots seem to make a line going up from left to right, it means that as one thing increases, the other tends to increase too. In our height-weight example, this would mean that taller people tend to weigh more.
- **Negative Correlation**: If the dots make a line going down from left to right, it means that as one thing increases, the other tends to decrease. For example, this might happen if we plotted "time spent studying" against "number of questions missed on a test".
- **No Correlation**: If the dots look randomly scattered with no clear pattern, it might mean there's no strong connection between the two things we're looking at.

## *Strength of Correlation*

Correlation strength is quantified by a coefficient that can range from (minus one) -1 (perfect negative correlation) to (plus one) +1 (perfect positive correlation). This coefficient is a numerical indication of the correlation between two variables.

The strength of correlation can be explained as follows:

- **Strong correlation**: coefficient values between ±0.7 and ±1.0 suggests that variables move linearly and predictably with one another.
- **Moderate correlation**: coefficients ranging between ±0.4 and ±0.7 suggest a significant but not necessarily controlling relationship.
- **Weak correlation**: coefficients ranging between ±0.1 and ±0.4 suggest a minor relationship.
- **No correlation**: coefficients ranging between -0.1 and +0.1 indicate no significant linear relationship.

In scatter plots, strong correlation manifests as a tight cluster of points that forms a straight or curved line, a lower degree of correlation manifests as a broader, more diffused pattern. Correlation strength tells us how significant the relationship is between two variables — whether the relationship is worth exploring or may simply be a coincidence.

## *Regression*

Regression analysis goes further than finding correlation: it constructs a mathematical model that tries to predict how the dependent variable varies when the independent variable changes. While correlation indicates whether two variables are associated, regression explains the nature of the association.

**Relationship of GDP Per Capita to Life Expectancy (2022)**

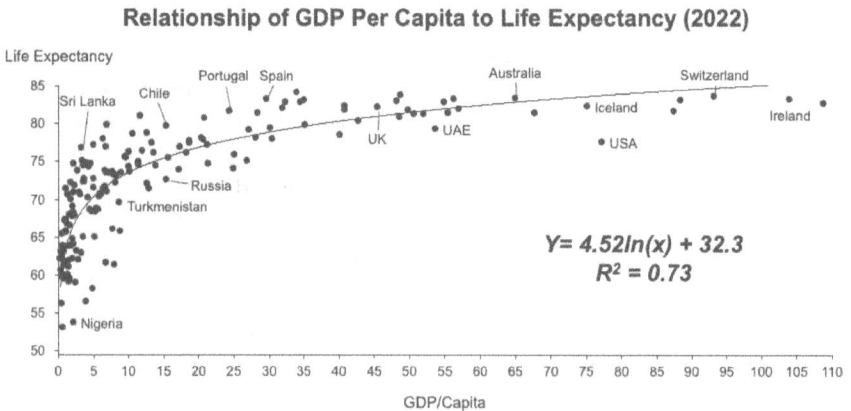

$$Y = 4.52\ln(x) + 32.3$$
$$R^2 = 0.73$$

Regression models can take many forms based on the pattern of the relationship, such as linear, logarithmic, exponential, polynomial, and so on. A regression model is usually evaluated based on R-squared ($R^2$)— the percentage of variance in the dependent variable that can be explained by the independent variable(s). A higher R-squared means a better fit.

The majority of data visualization tools (for example, Excel and PowerPoint) provide either default functions to compute regression models or to plot R-squared values on a scatter plot.

Consider the example of the relationship between GDP per capita and life expectancy, which tends to be logarithmic, not linear. As shown in the graph below, this relationship can be characterized by the following equation: $Y = 4.52\ln(x) + 32.3$.

The logarithmic model captures an important real-world pattern: as GDP per capita increases, life expectancy rises as well, though the effect starts to falter (diminishing returns) above about \$20,000 per capita. This means that beyond this threshold, more wealth yields smaller gains in life expectancy.

In this example, each data point would be represented by a dot on the chart, with the horizontal position determined by the advertising spend and the vertical position determined by the sales revenue. If the dots form a pattern that slopes upward from left to right, it indicates a positive correlation between advertising spend and sales revenue. Conversely, if the dots slope downward, it suggests a negative correlation. If there is no clear pattern, it implies no correlation between the two variables.

Scatter plots are most useful when:
- You have two sets of numbers to compare.
- You want to see if there's a relationship between these numbers.
- You have enough data points to see a pattern (usually at least 30).

They're commonly used in science, business, and social studies to look at things like:
- How temperature affects plant growth.
- The connection between advertising spending and sales.
- How study time relates to test scores.
- The relationship between a country's wealth (GDP per capita) and life expectancy.

# Pie charts: Visualizing Simple Proportions

Pie Charts are ubiquitous charts that divide a circle into sections representing proportions of a whole value, often used for simple breakdowns like product categories, traffic sources, or survey results. While they effectively show dominant versus non-dominant shares, people generally don't estimate pie wedge areas very well. More than a few slices can make values hard to distinguish and quantify. Pie Charts are one of many alternatives used in visualizing breakdowns, which include the donut charts, waterfall charts and stacked bar charts.

**Product Market Share**

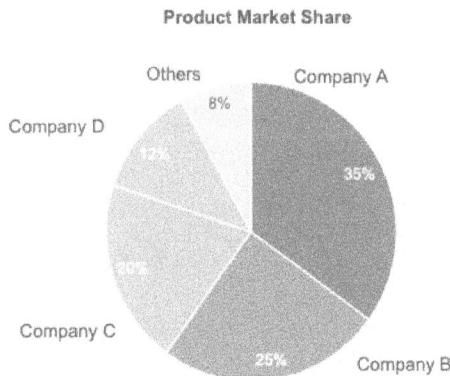

Key Features of Pie Charts:
- **Composition**: Pie charts are used to show how a whole is divided into different parts or categories. Each slice of the pie represents a category, and the size of the slice represents its proportion of the total.
- **Proportions**: The angles of the slices in a pie chart are proportional to the quantities they represent. The entire pie represents 100% of the total, and each slice represents a percentage of that total.
- **Categorical Data**: Pie charts are best suited for categorical data, where each category is distinct and mutually exclusive.
- **Limited Categories**: Pie charts are most effective when there are a limited number of categories (typically up to 6). Too many slices can make the chart cluttered and difficult to read.

When to use Pie Charts:
- There are a small number of segments in the pie.
- The difference between the segments is large and easy to see.
- There are no small segments (or they can be combined into an "Other" category).

- The overwhelming focus is on showing all the parts of the whole, rather than a comparison between segments.
- The audience is unused to seeing information graphically.

### Example: Market Share by Company

Let's say you want to visualize the market share of different companies in a specific industry.

In this example, a pie chart would have each company represented by a slice of the pie. The size of each slice would be proportional to the company's market share. The "Others" category combines the smaller companies to keep the chart readable.

# Donut Charts: Refining Proportions (and Adding Space)

Donut charts are often a better alternative to pie charts. While pies provide a full view of the surface area and the proportions are driven by angles, in a donut chart, the arc length is the intuitive driver of the breakdown. That said, while pie charts become messy if you have more than a few variables to visualize, the donut chart helps overcome this limitation by stacking two or more donuts.

Donut charts are a variation of pie charts that have a circular hole in the center, resembling a donut shape. Like pie charts, donut charts are used to visualize the composition or breakdown of a whole into its parts. However, donut charts have some advantages over traditional pie charts, particularly when dealing with multiple categories or when you want to emphasize specific information.

Key Features of Donut Charts:

1. Composition: Donut charts, like pie charts, show how a whole is divided into different parts or categories. Each segment of the donut represents a category, and the size of the segment represents its proportion of the total.
2. Central Space: The hole in the center of the donut chart provides space to display additional information, such as the total value, the title, or key metrics related to the data.

3. Multiple Levels: Donut charts can be used to display multiple levels of data by having concentric rings, each representing a different level of categorization. This allows for a more detailed breakdown of the data.

4. Emphasis on Key Categories: Donut charts can be effective in emphasizing specific categories by visually separating them from the rest of the data. This can be achieved by pulling out or exploding a segment of interest.

- **Example: Sales by Product Category and Sub-Category**: Let's say you want to visualize the sales breakdown by product category and sub-category.

Product Breakdown By Category and Sub-Category

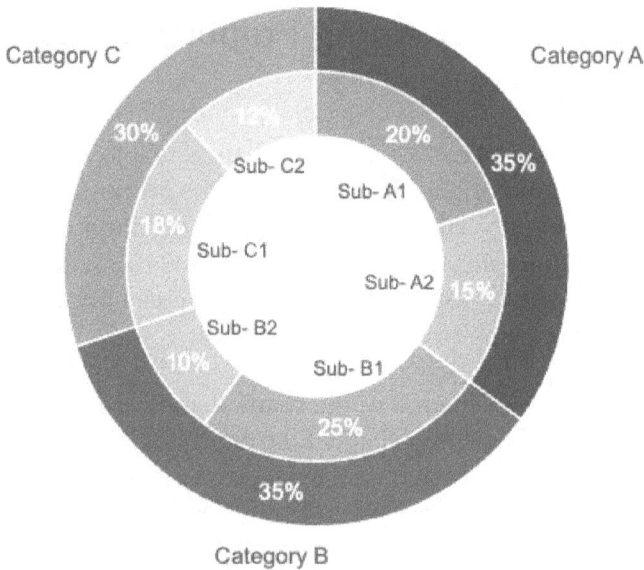

Category C                                 Category A

30%    12%    20%    35%

Sub- C2    Sub- A1

18%

Sub- C1    Sub- A2   15%

Sub- B2

10%    Sub- B1

25%

35%

Category B

In this example, a donut chart with two levels can be used to visualize the data. The outer ring would represent the main product categories (A, B, C), while the inner ring would represent the sub-categories within each main category. The size of each segment would be proportional to its sales percentage.

## *Donut Charts vs. Pie Charts*

Donut charts offer some advantages over traditional pie charts in certain scenarios:

1.  **Multiple Levels**: Donut charts can effectively display multiple levels of categorization by using concentric rings. This allows for a more detailed breakdown of the data compared to a single-level pie chart.
2.  **Central Space Utilization**: The hole in the center of the donut chart provides valuable space to display additional information, such as the total value or key metrics. This can enhance the overall information conveyed by the chart.
3.  **Emphasis on Key Categories**: Donut charts can be used to emphasize specific categories by visually separating them from the rest of the data. This can be achieved by pulling out or exploding a segment of interest, making it more prominent.

## *Considerations and Limitations*

Donut charts come with some limitations that you should consider when deciding whether to use them:

1.  **Perception of Angles and Areas**: Like pie charts, donut charts rely on the visual perception of angles and areas to represent the proportions of categories. This can make it challenging to accurately compare the relative sizes of the segments, especially when the differences are small.
2.  **Limited Data:** Donut charts, similar to pie charts, are not suitable for displaying large amounts of data or many categories. As the number of segments increases, the chart becomes cluttered and difficult to interpret.
3.  **Comparison Difficulty**: Comparing multiple donut charts side by side can be challenging, as the viewer needs to mentally compare the sizes of the segments across different charts.

In summary, donut charts are a useful alternative to pie charts, particularly when you want to display multiple levels of categorization or emphasize specific categories. The central space in the donut chart can be utilized to provide additional information or key metrics. However, like pie charts, donut charts have limitations in terms of accurate comparisons and handling large amounts of data. By understanding the strengths and limitations of donut charts, you can make informed decisions on when to use them effectively in your data storytelling.

# Mekko Charts: Seeing the Big Picture and the Details

Now let's pivot to discuss the chart type that may seem more complex but packs a serious punch when you need it: the Marimekko Chart (or just Mekko for short). Another name it goes by is variable width chart or mosaic plot, or you might even see it appear in top-level business presentations without a name – its unique appearance gives it away.

Imagine a stacked bar chart where one bar equals one total broken down into components. Now, what if the *width* of each of those bars also represented something meaningful? What if wider bars represented larger categories in aggregate? This is the basic concept of a Marimekko. It allows you to view two categorical dimensions simultaneously:

- **The Width**: The width of each stacked bar represents the absolute amount of that category (i.e., total market share of a company, total sales in a region).
- **The Height:** In each bar, the height of the segments indicates how much of that second categorical variable is represented (e.g. product category mix of that company's share, revenue sources of that region's sales).

## *Why bother with the additional complexity?*

So, why do this instead of a simpler stacked bar or two pie charts? Because it allows you to see relationships and compositions that would otherwise be hidden or require multiple separate charts to piece them together.

- **Beyond Simple Stacking**: While a simple stacked bar chart allows such comparisons between the internal composition of your categories, it doesn't show the size difference between them overall. All the bars are the same width. The Marimekko solves this problem — you immediately see which bars (categories) are the largest overall while simultaneously showing the breakdown within each category.
- **Two Variables, One View**: It is uniquely able to decompose how composition changes across categories of different sizes. For instance, do our largest customers (wide bars) purchase a different mix of

products (segment heights) than our smallest customers (narrow bars)? A Marimekko chart can provide this insight visually.
- **Visual Pattern Recognition**: With variable widths, different heights of segments, and color-coding, the elements create a visual tapestry. Your eyes can rapidly identify dominant segments in big categories, register deviations (e.g., a small category with an outlandish internal mix) and sense the overall skeleton of the data in a way impossible to achieve with other chart types.

## *Example: Sales by Product & Region*

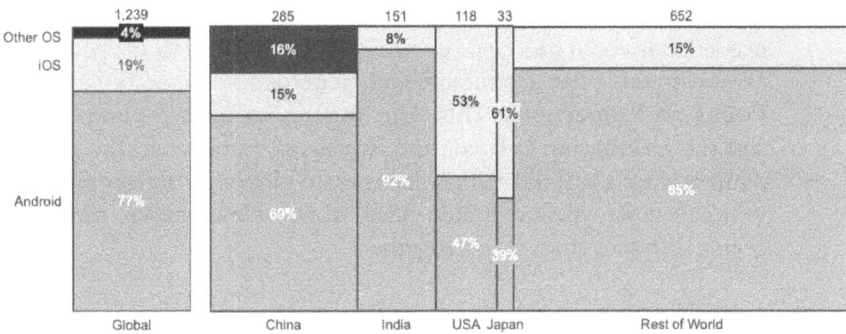

**Global Smartphone Sales Breakdown**
In Million Units Sold, 2024

For example, consider smartphone sales by country (China, India, East, USA, Japan, and Rest of the World) and product categories (iOS, Android, Other).
- In a Marimekko, each country would be a bar. The bar's width would represent that country's total unit sales volume.
- Inside each of the country bars, the height of the segments would show the share of iOS, Android, and other OS in that country.

So, now if you were analyzing Apple iPhone sales because you are considering launching a new smartphone application but want to decide which OS platform you should be prioritizing, you can answer questions such as:
- What would be your potential global market reach from new smartphone buyers?
- Where would the target market be coming from in the largest markets vs. rest of the world?

- If the USA is your primary target market, then which OS should be prioritized for the widest reach?

## *Here's the Catch (Considerations & Limitations)*

Marimekko charts have many strengths, but they're not the right tool for every situation.

- **Complexity Can Bite**: If you add too many bars (regions) or too many segments (products), suddenly, the chart becomes a mess. Most of all, you need to balance detail with readability. Less is often more.
- **Judging Areas Visually is Difficult:** It is challenging for the human eye to accurately compare the area of various differently colored rectangles (especially rectangles of different shapes)—just like with pie charts. If accurate observation for comparison between segments is needed, this could be a poor choice without clear labels. It's better for demonstrating relative structure and patterns.
- **Focus on Proportions**: This chart focuses on relative composition and the relationship between the two items, rather than the precise numerical values. If the main objective is to pinpoint accuracy on every value, then yes, you'll definitely need labels or annotations, or maybe even a different chart type altogether.

## *The Bottom Line*

If the narrative's main point isn't just the parts of a whole but the differences among those parts across wholes of varying size, then by all means, reach for a Marimekko chart. Mekko charts are great to visualize the relationship between two categorical variables (and composition) in the context of overall scale. They offer a uniquely insightful visual perspective that simpler charts often can't. Only make sure to use it judiciously and design it cleanly to prevent your audience from being overwhelmed.

# Barbell Charts

Barbell charts are a unique type of visualization that has gained popularity with prominent financial publications such as The Economist and Financial Times. Characterized by the lateral line that connects two different points and resembles weightlifting equipment, this chart powerfully showcases the change or gap between two data points for a single category and enables the viewer to easily compare the change between two states.

Unlike bar or column charts, which emphasize absolute values, the barbell chart's superpower lies in its emphasis on the distance and direction of the movement. This feature makes barbell charts most suited for "before" and "after" comparisons, shifts over time, or comparing actual results to target values.

## How Does it Work Visually?

The format is crisp and laser-focused:

- **Two Markers:** Typically circles or dots, denoting the values at the two points to be compared.
- **Connected Line**: Draws a straight connecting line between the two markers.
- **Category Label**: Next to the barbell, specifying what is being measured.
- **Value Labels (optional but recommended):** Usually found right next to the markers to indicate their respective values.
- **Sorting**: Categories are often sorted by size or direction of change to reveal and emphasize a specific pattern.

## Why Use Barbell Charts?

Because these kinds of charts are a relatively a new visualization technique, some elements of their design may be confusing in some regards. However, this type of visualization has become increasingly popular with top financial publications for several compelling reasons. They enable viewers to slice through noise and immediately deliver the heart of the story – investors gauging changes, comparisons between two points in time (before/after) – which is valued. The Economist, for example, uses a barbell chart to compare GDP change between different countries, shifts in market indices, or fluctuation in currency during a certain period.

They enable the audience to quickly understand:

- What categories saw the most and least changes.
- The nature of the change (up or down).
- The size of the difference between the two positions.

## *Best Practices:*

If you decide to design your own barbell chart, consider these best practices in your design:

- **Keep it Clean**: Avoid excessive clutter, as the barbell's strength lies in its simplicity.
- **Sort Meaningfully**: Compare categories side by side, based on either the change amount or the value at the end.
- **Use Color with a Purpose:** If you are using color on the markers or line, consider using color to represent the direction of change (i.e., green when there is an increase, red when there is a decrease) or to isolate certain categories.
- **Use Clear Labels**: Use easy-to-read categories and values (if displayed), e.g., Indicate accurate values on both ends of the chart.
- **Limit Categories**: As is the case with most charts, too many barbells can make it hard to process the information. Concentrate on the most applicable comparisons.
- **Comparison Marker:** Add a secondary marker for benchmark or average values.

## *Example: Change Over Time*

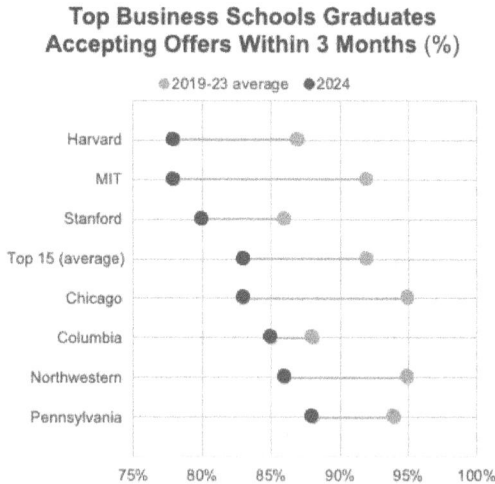

**Top Business Schools Graduates**
**Accepting Offers Within 3 Months (%)**

● 2019-23 average  ● 2024

| | 75% | 80% | 85% | 90% | 95% | 100% |
|---|---|---|---|---|---|---|
| Harvard | | | | | | |
| MIT | | | | | | |
| Stanford | | | | | | |
| Top 15 (average) | | | | | | |
| Chicago | | | | | | |
| Columbia | | | | | | |
| Northwestern | | | | | | |
| Pennsylvania | | | | | | |

The barbell chart that follows illustrates the change in job offers acceptance rates for top business school graduates between 2019-23 (average) and 2024 — and how well the chart format does at highlighting comparative performance. Each school is represented by a row with two-shaded red circles (the two periods) connected by a horizontal line whose length instantly conveys how much change occurred. At a glance, the chart shows a few important things: MIT and Harvard had the most dramatic drop, while Columbia, Pennsylvania, and Standford dipped slightly, and most schools experienced a 10-20% drop. The clean grid lines provide context for how precise the measurements are without unnecessary visual distraction, allowing viewers to easily identify trends and make informed comparisons across many institutions at once—which is why publications like The Economist and Financial Times often prefer to use this chart when comparing comparative changes.

## The Bottom Line

The beauty of the barbell chart is its simplicity. By eliminating all but the most basic elements of change, all that is left is the transformation between two points in time which can, therefore, achieve an immediate visual impact that is often hard to equal when using more complex chart types. This is why publications such as the Financial Times, which are known for their excellence in data visualization, have adopted this format as a staple in their visual communication arsenal.

If your story is about the difference, change, or transition between two points in time for several variables, then the barbell chart is a powerful, easy-to-read way to get your point across. This singularity directs focus straight to the scale and direction of change, proving itself as an invaluable tool in the data storyteller's toolset.

# Slope Charts

Similar in spirit to the barbell chart but used for slightly different nuances more often, the Slope Chart is another really effective visual for showing change over time, especially rank shifts among categories. Popularized by the visualization expert Edward Tufte and commonly found in magazines like The Economist, slope charts reduce trends to their most basic elements: the start point, the end point, and the slope that connects them. They take the richness of many data points and reduce them to a relationship between one starting and one ending value. The Financial Times often uses them to demonstrate changes in economic indicators between countries, changes in performance across sectors, or changes in key metrics between reporting periods.

## *How Does it Work Visually?*

They're like line charts (but less cluttered), and instead of plotting many intermediate points, they just focus on the value for each category at each of the two specific points in time (or under two conditions) you care about. The basic skeleton of a slope chart consists of:

- **Two Parallel Axes (optional)**: Represent the two points in time (e.g., 2015 and 2020) being compared.
- **Connecting lines**: For each category, you have a connecting line between the value on the left-axis and the value on the right-axis.
- **Slope Signifies Change**: The slope or angle of the line, immediately conveys:
  - o **Direction**: Up means increase, down means decrease.
  - o **Magnitude** is the amount of change: A steeper slope means a larger change.
- **Rank Changes**: When lines cross, we can clearly see the categories experiencing rank changes.
- **Labels**: The category names and values are typically printed directly on the ends of the lines.

## *Why Use Slope Charts?*

They are powerful because they expose many dimensions of change across many categories at once:

1. Direction of change (upward or downward slopes)
2. Magnitude of change (steepness of slopes)
3. Ranking shifts (lines that cross indicate changes in relative position)
4. Outliers (unusually steep or contrary slopes)

## *Best Practices*

Use these best practices when creating slope charts:

1. Limit the number of categories to avoid visual clutter (typically 5-15 lines).
2. Use color strategically to highlight important shifts or group related categories.
3. Consider sorting the data points on one or both axes by value.
4. Ensure adequate spacing between lines to maintain readability.

5.  Add contextual annotations to explain significant changes or anomalies.

## Example: The Magnificent Five

The example here shows the change in market capitalization for prominent tech stocks from 2015 to 2020; it shows the power of the format in visualizing growth paths. The chart places the time periods in time on two vertical axes (2015 is on the left, 2020 is on the right) with diagonal lines connecting each company's values, immediately communicating both the degree and direction of change. The slope of each line speaks to its rate of growth — NVIDIA's rapid angle demonstrates its skyrocketing ascent from $0.01T to $3.33T, while Exxon Mobile's more horizontal line shows a much more modest pace. Both endpoints have data labels that give specific values, so there is no need to refer to any axes. The use of a blue hue and varying opacity allows for consistent coloring in the chart while enabling viewers to differentiate between companies. This is a classic case of why a slope chart should be used for comparative performance across multiple entities over two separate time points (financial report cards, competitive comparisons, and trends).

### Market Cap Change for Selected Stocks

Trillions

Apple, $3.58
NVIDIA, $3.33
Microsoft, $3.11

Alphabet, $2.36

Meta, $1.55

Apple, $0.59
Alphabet, $0.53
Microsoft, $0.44
Exxon Mobile, $0.36
Meta, $0.30
NVIDIA, $0.01

Exxon Mobile, $0.47

2015     2020

## The Bottom Line

Slope charts work very well when your story is focused on:
*   Comparative performance among categories
*   Ranking changes over time

- Recognizing winners and losers in a relative sense
- Emphasizing convergence or divergence tendencies

The simplicity of slope charts makes them a perfect complement to more complex visuals in larger presentations. An accompanying slope chart can further emphasize the before-and-after part of the story that is often more important to the decision-making process, even if a detailed line chart is available showing the entire trajectory.

When creating your presentations, opt for slope charts to visualize comparative change across categories. The combination of visual simplicity and ease of interpretation makes them particularly useful for audiences who are intellectually curious and who crave a combination of analytical clarity and design elegance.

# Bullet Charts

Developed by data visualization expert Stephen Few as a more informative and space-efficient alternative to dashboard gauges and meters, the **Bullet Chart** is a powerhouse for displaying **performance against targets** within context. It packs a surprising amount of information into a compact linear format, making it ideal for dashboards and performance summaries.

Think of it as a bar chart on steroids. While a simple bar shows a single value, the bullet chart layers that value with comparative information and qualitative context.

## *How Does it Work Visually?*

Unlike simple bar charts, which are a simplistic depiction of raw numbers in isolation, bullet charts provide more contextual information by embedding multiple Key Performance Indicators (KPIs) within a holistic visual framework. A bullet chart consists of:

- **Primary Measure:** The main data point (e.g., year-to-date revenue) is displayed as a central bar (the "bullet").
- **Comparative Measure:** A target or benchmark value (e.g., the revenue target) is shown as a short perpendicular line or marker.

- **Qualitative Ranges:** Background shading indicates performance zones (e.g., poor, satisfactory, good). These provide immediate qualitative context.
- **Scale:** A clear quantitative axis runs alongside the chart.
- **Label:** Identifies the metric being displayed.

## *Why Use Bullet Charts?*

Bullet charts illustrate analytical power because they allow you to answer multiple evaluation questions all in one compact visual. For example, when displaying performance, they help answer the following questions:

1. What is the current performance level? (The length of the main bar)
2. How does it compare to the target or benchmark? (Position relative to the marker)
3. Where does it fall within defined performance ranges? Is performance satisfactory, excellent, or concerning? (Which background zone does the bar reach?)

This density of information makes them extremely efficient for conveying performance status quickly and clearly, a reason they are favored in business intelligence dashboards and reports.

## *Best Practices:*

When creating bullet charts, make sure to adopt the following best practices:

- **Keep Ranges Simple:** Usually, 3-5 distinct qualitative ranges (shades) are sufficient. Too many becomes confusing.
- **Use Intuitive Shading:** Typically, darker shades represent better performance zones but ensure consistency and clarity.
- **Clear Target Marker:** Make the comparative measure visually distinct.
- **Consistent Scales:** When comparing multiple bullet charts vertically (e.g., for different regions or KPIs), use consistent scales if feasible to allow for accurate visual comparison.
- **Label Clearly:** Ensure the metric, scale, and target value are unambiguous.

## Example: ACSI for Smartphones

The bullet chart that follows displays the American Customer Satisfaction Index (ACSI) for the most popular smartphone manufacturers. Each manufacturer is represented by a horizontal bullet chart, with the blue bar showing the 2024 score and the red vertical marker showing the 2023 score. The chart clearly shows that the two leaders, Apple and Samsung, are tied with a score of 82, both showing improvement over the previous year. The industry average is 79, while Google and Motorola are below average, with a score of 77 each.

**American Customer Satisfaction Index (ACSI)**
**for Smartphones**

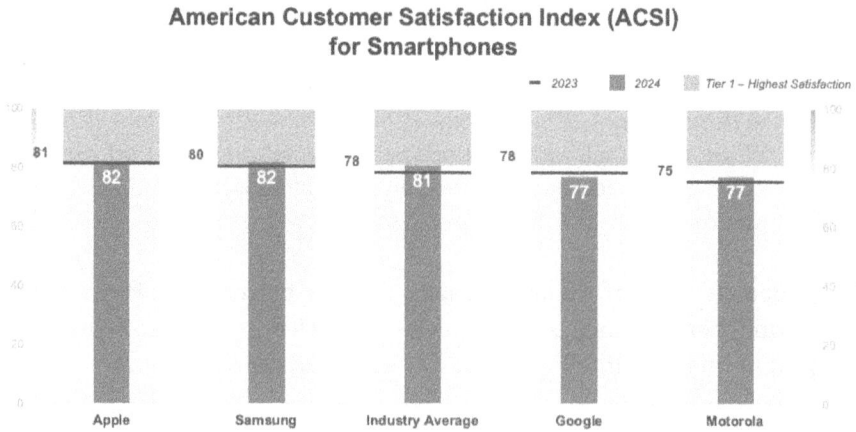

Complementing the charts' scores is a qualitative background horizontal bar. It makes it immediately clear which manufacturers exceed or meet the industry baseline. Similarly, comparative markers (previous year's scores) help viewers quickly contextualize each manufacturer's year-over-year performance changes. Through a single, compact visualization, we can see not only where current performance stands but also relative positions against industry benchmarks, demonstrating the bullet chart's exceptional ability to convey multiple layers of performance data in an easily digestible format.

## The Bottom Line

Bullet charts truly shine when in your presentation needs to:
- Show progress against goals
- Benchmark actual performance
- Provide contextual reference points for interpretation

- Condense multiple related performance metrics in a single visual

Bullet charts are the go-to choice when you need to display a primary measure alongside its target and qualitative performance context in a minimal amount of space. They offer unparalleled information density for performance monitoring and reporting, making complex assessments intuitive and immediate.

The high information density of bullet charts makes them an invaluable component of executive dashboards and summary slides, where space is limited but a rich analytical context is required. Whereas a standard bar chart might convey just absolute values, the bullet chart conveys value, target, and performance thresholds all at once—and all in a single, coherent visual.

In creating your presentations, consider using bullet charts when the story is about performance or benchmarks. Their visual compactness combined with contextual density makes them especially powerful for complex audiences who need to rapidly assess performances on multiple dimensions without compromising analytical depth.

# Waffle and Plum Charts

Waffle charts and their stylistic variant, known as "plum charts," are distinctive visualization approaches that have become signature elements in McKinsey & Company's visual communication toolkit. These charts elegantly meet the challenge of presenting part-to-whole relationships, most notably when mapping onto percentages or proportions.

The waffle chart, also referred to as a square pie chart, takes the form of a grid of squares (usually 10×10) with each individual square representing a proportional fraction of the whole (1%). Coloring these individual squares according to different categories gives a visual that shares the part-to-whole clarity of pie charts and the precision of a grid system.

The "plum" variant, commonly used in McKinsey visuals, utilizes circles instead of squares and sometimes introduces additional design embellishments for visual improvement without compromising analytical integrity.

## How Does it Work Visually?

Waffle and plum charts typically include:
- **Grid Layout**: Usually a square grid (e.g. 10x10 units = 100 units).
- **Unit Representation**: Precise proportional representation where each element equals a specific percentage. For example, the circle or the unit square represents an equal fraction of the whole (or some specific number, e.g., 1%, 1,000 people).
- **Color Coding**: The various categories (e.g. quarters and types) have different colors assigned to them, the equivalent amount of each unit is colored in the same color, and clear labeling indicate the categories and their percentage values.
- **Clear Proportions**: The number of colored units is directly related to the proportion of each category.
- Often, a total count or value is included to provide scale.

## Why Use Waffle/Plum Charts?

These charts are adopted by expert presenters due to their ability to communicate a proportional idea with breathtaking precision and visual impact. This makes them more visually intuitive than common pie charts, which can sometimes trick us into thinking one portion is larger or smaller than it is, while waffle and plum charts lend themselves to better proportion judgement:

1. **Easier Comparison**: Humans can compare areas made of discrete units with greater accuracy than comparing angles or arc lengths in pies or donuts.

2. **Precision**: They allow for precise counting of individual units when needed.

3. **Visual Engagement**: The grid format is potentially more visually striking and less likely to be distorted (unlike pie charts).

4. **Consistency**: They maintain a consistent shape regardless of the data, making comparisons across multiple charts more reliable

5. **Flexibility**: You can change the size of the grid or unit meaning and apply it to other total numbers.

6. **Categorization**: They can effectively display multiple levels of categorization through nesting or grouping.

## *Best Practices:*

Here are some best practices for creating waffle or plum charts:

- **Maintain Consistent Grids**: For side-by-side analysis with multiple waffle charts, keep grid sizes the same (typically 10×10 for simplicity) for fair comparison.
- **Logical Arrangement**: You may want to keep the colored blocks close together for each category so it will be easier for counting and perception. Commonly, begin input in the same corner (i.e. top-left or bottom-left)
- **Clear Labeling**: Use clear labels and a legend that maps colors to categories showing what & how much each category represents. Consider adding percentage labels directly on the chart for key categories.
- **Coherent Color**: Use a coherent color palette that differentiates categories while maintaining visual harmony.
- **Limit Categories (Sometimes):** Even though they do better than pies with many categories, too many small categories can make a plot seem fragmented. If necessary, include an "Other" category.
- **Accessibility**: Make sure the colors for your categories are contrasting enough.

## *Example: The 80/20 of US Foreign Aid in 2023*

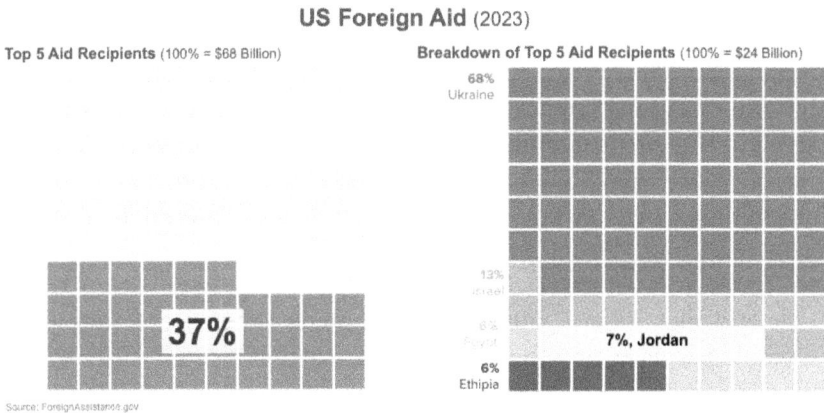

**US Foreign Aid** (2023)

Top 5 Aid Recipients (100% = $68 Billion)

Breakdown of Top 5 Aid Recipients (100% = $24 Billion)

68% Ukraine

13% Israel

7%, Jordan

6% Ethipia

37%

Source: ForeignAssistance.gov

Using a grid of small squares, the following waffle chart conveys how U.S. foreign aid was allocated in 2023, with each small square representing a uniform percentage of the whole. The left grid shows how 37% of aid is concentrated in the top 5 recipients, while the right grid breaks down this 37% among those countries (with Israel receiving the largest portion at 16%, followed by Ukraine, Egypt, Jordan, and Iraq). There is color-coding used for countries, and percentages are clearly labeled and consistent. One thing you can do with the square formatting is make sure the value of the parts-to-whole comparisons are represented by tight grid distances apart, something that would be difficult to ascertain in a pie chart, especially given the (nesting below) hierarchy between the total aid budget and the breakdown of its top recipients. Interestingly, McKinsey consultants love the waffle chart for hierarchical proportional data — this looks amazing as compared to other charts because it shows the aggregate value with the packing efficiency of the grid for counting.

## *The Bottom Line*

These charts are especially effective when you need to:
- Communicate proportional relationships with precision
- Display categorical breakdowns in a visually engaging manner
- Show multiple comparative distributions side-by-side
- Balance analytical accuracy with design elegance

McKinsey's uptake of these charts Signals the firm's embrace of communication that is both disciplined and visual. Now, a pie chart works fine for simple proportions, but waffle and plum charts give us a more precise proportion representation while remaining visually appealing. So, for more nuanced, proportional comparisons that require precision and visual impact in equal measure, consider waffle or plum charts in your presentations. Their grid-based structure coupled with intuitive color-coding make them exceptionally useful for sophisticated business audiences who seek analytical rigor alongside polished presentation.

Waffle and Plum charts work great to visualize the exact composition of a whole partwise, presenting them in a way that is often easier to read than pie or donut charts. Their grid-based structure lends itself to clarity and visual appeal, making them effective for representing steps out of the total, questionnaire results, or resource allocations, especially for viewers who prefer a straight-to-the-point approach and few bells and whistles.

# Pictograms and Isotopes Charts

Lastly, let's look into a technique that brings a layer of intuitive meaning and interactivity to almost any chart type: Pictogram Charts (often referenced more formally by their historical system name, Isotype — International System of Typographic Picture Education). Rather than using bars or dots or other abstract shapes to communicate the quantity of data points, these charts use relevant icons or symbols.

Pictogram and Isotype charts are a flexible visualization method that can turn almost any chart into a better one, as they include recognizable icons or symbols to illustrate categories and values. We are so conditioned to see graphs as arbitrary shapes, lines, and dots that replacing the standard shapes with meaningful icons directly connects the chart to the subject matter, thus making the data and content more relatable and memorable.

What distinguishes pictogram charts from anything else we've covered so far is their ability to take the iconic approach and apply it to (almost) any charting framework discussed in this chapter.

Pictogram charts trace their origins to the Picture Language (International System of Typographic Picture Education) that Otto and Marie Neurath created in the 1920s. Their objective was to create a universal visual language for communicating statistical information across literacy and language barriers. Fast-forward to today, this methodology has transformed into advanced data visualization methods that maintain the human-centric essence of the original idea.

## *How Does it Work Visually?*

Pictographic elements can enhance bar charts, column charts, waffle charts, area charts, and even line charts to make them more engaging and persistent.

The fundamental structure of pictogram charts includes:
- **Icons, Icons and More Icons**: Subject-relevant icons that replace or enhance standard chart elements.

- **Descriptive Variation:** Simple and standard symbols specific to the data (e.g., figures for population, cars for car sales, trees for deforestation).
- **Proportional Sizing:** Consistent sizing and spacing of icons to maintain quantitative accuracy.
- **Clear Indication of Scale:** (e.g., "each car icon represents 1 million vehicles"). Per Unit Value — Understand what amount the icon represents: An icon usually represents a certain quantity of an entity (1 icon = 1 million people, for instance). The icon is repeated to display the total amount.
- **Legend and Labeling** are necessary to clarify what amount of something an individual icon contains.
- **Integration with Other Chart Types:** Pictograms themselves are not a stand-alone chart type but rather an application of the technique to other charts. Here are some variants to consider when using pictogram charts:
  - **Unit Charts:** Where each icon represents a specific quantity and is repeated to show the total (similar to waffle charts but using themed icons).
  - **Icon-Enhanced Bar/Column Charts:** Where standard bars are either replaced by rows of icons or enhanced with relevant imagery.
  - **Proportional Icon Charts:** Where the size of a single icon changes to represent different values (though this approach requires careful implementation to avoid distortion).
  - **Icon-Infused Area Charts:** Where pattern fills using small icons replace solid colors in area charts.

## Why Use Pictograms?

Their key advantage is linking abstract numbers to flesh-and-blood meaning. The strategic benefit of pictogram charts becomes clear from their ability to create immediate cognitive connections.

- **More Relatable:** Icons help ground the data, making it that much less abstract, and more in touch with the subject.
- **Better Retention:** Symbols are typically more memorable than plain bars or digits. They often increase retention of the information presented.
- **Improved Engagement:** They transform abstract numbers into visually tangible elements, which can help make charts more visually attractive and interesting.

- **Accessibility:** Can facilitate understanding across language or literacy barriers (the original aim of Isotype). They can transcend language barriers in international presentations.
- **Cognitive Ease:** They reduce the cognitive load required to understand the subject matter.

## Best Practices

When creating pictogram charts, follow these best practices:

1. **Stick to Clear, Simple Icons** — Icons should be easily recognizable and not leave room for interpretation. Do not use too complex or detailed images.
2. **Icon Size**: Maintain consistent icon sizing and spacing to preserve quantitative accuracy. Crucially, each icon should represent the same thing; this is critical to support the proportion. When representing value, never scale the size of icons to indicate value, as this tricks perception. Instead, use more icons of the same size. If you must, fractional icons may be used sparingly and in context.
3. **Maintain Consistency**: If you are using the same icons when creating a series of related charts, use icon that are of the same style, size, etc. Provide a clear key: Always state clearly what one icon stands for and provide clear scale references.
4. **Don't Go Too Far**: These visual markers should facilitate comprehension, not overwhelm or distract from the central premise. Balance visual interest with data integrity – the icons should enhance, not obscure, the data story.

## Example: The Wildfires of California

As a quick reminder, don't forget to download the companion visual booklet at www.storytellingwithcharts.com/vl-stcai.

The two visualizations work well together in showing California's largest wildfires through complementary approaches: one as a bar chart with flame icons, and the other as waffle pictograms in which each flame icon represents 10,000 acres burned.

In the bar chart version, we can see the dramatic scale differences between the major fires. The August Complex fire (2020) and Dixie fire (2021) dominate with their massive ~1-million-acre burns, creating bars that extend dramatically longer than the others. The subsequent fires—

Mendocino Complex, SCU Lightning Complex, Creek, and others—show a progressive decrease in affected areas, with names and exact acreage clearly labeled alongside each bar.

The second version transforms that same data into waffle-pattern flame icons, where each icon = 10,000 acres burned. This creates an immediate visual impact that connects viewers directly to the subject matter. The August Complex fire, which is pictured here with 102 flame icons (indicating its 1,032,648 acres), is dramatically larger than some of the smaller fires at the bottom of the chart. The precise, countable nature of the icons allows viewers to immediately grasp both the absolute and relative scale of each disaster.

**California 10 Largest Wildfires**
(2017-2024, In '000 Acres)

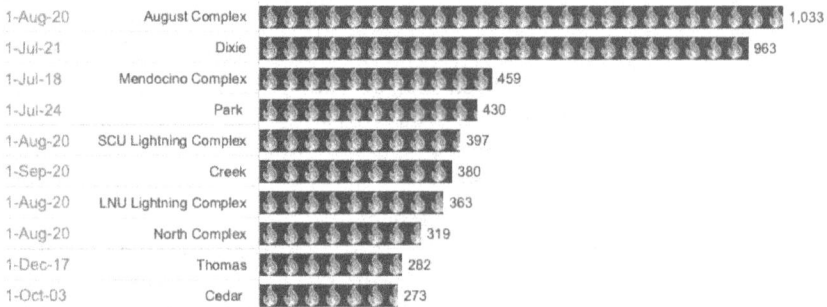

| Date | Fire | Acres |
|---|---|---|
| 1-Aug-20 | August Complex | 1,033 |
| 1-Jul-21 | Dixie | 963 |
| 1-Jul-18 | Mendocino Complex | 459 |
| 1-Jul-24 | Park | 430 |
| 1-Aug-20 | SCU Lightning Complex | 397 |
| 1-Sep-20 | Creek | 380 |
| 1-Aug-20 | LNU Lightning Complex | 363 |
| 1-Aug-20 | North Complex | 319 |
| 1-Dec-17 | Thomas | 282 |
| 1-Oct-03 | Cedar | 273 |

Seven out of the eight largest wildfires in California occurred in the last four years

**California 8 Largest Wildfires** (2008-2024, In '000 Acres)    ~10,000 acres

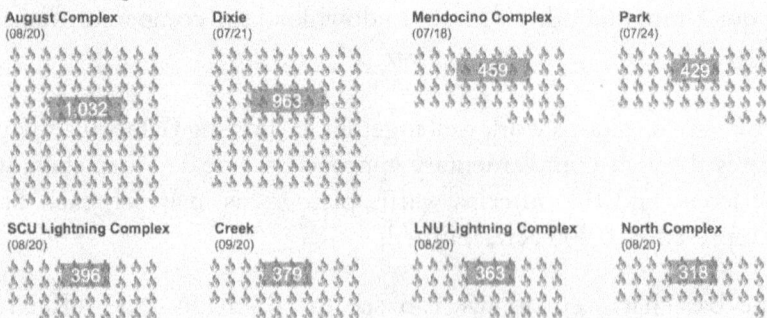

**August Complex**
(08/20)
1,032

**Dixie**
(07/21)
963

**Mendocino Complex**
(07/18)
459

**Park**
(07/24)
429

**SCU Lightning Complex**
(08/20)
396

**Creek**
(09/20)
379

**LNU Lightning Complex**
(08/20)
363

**North Complex**
(08/20)
318

Both visualizations illustrate how pictogram charts turbocharge data storytelling by substituting abstract bars with relevant imagery about subjects. This grounds the numerical data (acres burned) in a visual metaphor (flames) that creates instant recognition and emotional connection to the real-world impact of these wildfires, while maintaining the analytical clarity of the original bar chart format.

## *The Bottom Line*

Pictogram charts are particularly effective when your presentation aims to:

- Make statistical information more approachable and memorable
- Connect data to its real-world context
- Present information to diverse or non-technical audiences
- Create visually distinctive materials that stand out in a crowded information environment

Pictogram charts are a great way to keep some analytic rigor in your data story — and bring a notch of intuition that can sometimes go a long way in engaging audiences with your data. They are particularly great for executive summaries, public-facing reports, and circumstances where you want to quickly convey not just the numbers but the human context behind them.

When developing your presentations, consider how pictogram enhancements might elevate your standard charts into more memorable and engaging visual narratives. Their ability to bridge the gap between data and meaning makes them valuable tools in any data storyteller's toolkit.

# Chapter 5: Next Steps and Resources

You are on your way to mastering the art of storytelling with charts, and the principles you have learned will accelerate that journey. I look forward to seeing how you apply these techniques to tell data-driven stories that achieve successful outcomes.

As we conclude this book — a deep dive into the fundamentals of Vertical Logic — recall that we began with the notion that simply asking, "Given this data, what can I visualize?" doesn't work. And in a world overwhelmed with misconceptions hidden in data — misconceptions that need a clear, persuasive presentation to address — developing these skills is crucial to help you inform, persuade, and drive action.

In the previous sections, we learned how principles such as those from Gestalt psychology can help us create visual representations that will resonate naturally with our brains. And, more importantly, we've introduced the ultimate test: the "So What?" question, to ensure that each chart we create is directly relevant to our core message and strategic objectives.

I have deliberately kept this book focused on vertical logic, leaving the discussion of horizontal logic for my other book, "Storytelling with Charts: The Full Story". In the process, the intent was, for each slide to be created as a self-contained, powerful visual, regardless of context. This is fundamental, to master this framework of deck creation. It's all about maintaining clarity, usability, and credibility, one point and one slide at a time. It is the art and science of persuasion.

## The PowerPoint Advantage

In this book, we have been using Microsoft PowerPoint as our platform of choice for implementing storytelling charts. This choice wasn't arbitrary. PowerPoint is the industry standard in business presentations for a reason: it has powerful features, offers broad compatibility, and uses a familiar interface, making it relatively easy to learn more advanced techniques.

If you have a recent release of PowerPoint (especially Office 365) available, you could use it to create all of the visualization techniques presented in this book—its charting capabilities are extensive. With its widespread use, the platform also makes it less likely that your presentations will render incorrectly across different organizations or devices versions, which is essential when your aim is persuasive communication.

Alternative platforms work well for special use cases, but PowerPoint provides the optimal balance of capability, accessibility, and practicality for data storytelling with charts. The techniques you've learned can be applied most efficiently within this environment, especially when enhanced with purpose-built tools.

## Making PowerPoint Work for You

To accelerate your application of the principles covered in this book, consider exploring the Storytelling Charts Add-In (STCAI). This free tool integrates seamlessly into PowerPoint to significantly reduce the effort required for creating the types of charts and visual structure we have discussed.

This add-in includes templates specifically for implementing vertical logic, improved workflows for charting, and enhanced annotation capabilities—all tailored to the methodologies described in preceding chapters.

For readers wanting to put these ideas to work today, visit storytellingwithcharts.com. The tool is intuitive, and learning how to use it will pay off in time savings and improved presentation quality.

## The Enduring Value of Visual Storytelling

The principles we've discussed are enduring. Much like the longevity of ideas behind the Lindy Effect we talked about earlier, the essentials of *storytelling with charts* — structure, clarity, relevance — haven't changed significantly since their inception decades ago, and they are unlikely to change drastically anytime soon. They operate based on the way human cognition works. As you apply these principles, you become more than just the person who shows the data or writes a report; you become someone who explains it, who translates complexity into clarity, and who helps guide effective decision making. Such influence is invaluable in any field.

## How to Focus on the Bigger Picture Behind the Slide

As you master vertical logic, keep in mind that each powerful slide is part of a greater narrative – the Horizontal Logic that undergirds the entirety of your presentation or report. How do the messages on individual slides sequence together? In what order does the entire story unfold? How do you identify your speaking points and structure supporting evidence in a logical way?

If you're confident at crafting the individual chart but need to learn how it fits into a coherent, persuasive, overarching story, I urge you to take a look at the companion volume — Storytelling with Charts: The Full Story. It delves into more advanced horizontal logic techniques to give you the complete picture, from beginning to end, including a complete system for creating impactful presentations.

## Making It Real: Practice, Tools, and Continued Growth

Knowing something is powerful, but true mastery is only attained through applying that knowledge. The biggest benefits come when you begin practicing these principles, starting ***today***.

1. **Review Your Work**: Find a recent presentation. Choose one or two of your slides and reinvent them according to the frameworks and best practices we have discussed. You'll probably notice immediate areas for improvement.
2. **Build Your Own Library**: Begin building your own library of preferred chart templates in PowerPoint – templates with charts

set up using these principles. This will help automate your process over the long run.

3. **Seek Feedback**: Share your redesigned slides or new charts with colleagues you trust. Ask them: *What's your biggest takeaway? How long did it take for you to get it?* Honest feedback is invaluable for growth.

4. ***Learn from Experts***: *Study the work of top consulting firms, esteemed publications (The* Economist or the Financial Times*),* and effective communicators in your domain. Pay attention to how they visually organize their information.

To bridge the gap between the concepts in this book and your daily workflow, we have also developed the Storytelling Charts Add-In (STCAI) for PowerPoint. It embeds directly in the software you probably use every day, offering templates, efficient workflows, and annotation tools that are built based on the vertical logic principles we've discussed. It's designed to make it easier to implement what you've learned on an ongoing basis. You can find the Add-In and other resources, including information on more structured training programs that cover both vertical and horizontal logic, at storytellingwithcharts.com.

# A Final Thought

There is more to mastering visual storytelling — it's a lifelong journey — but you now have a solid starting point and a strong framework to create charts that actually tell a story! You know the reasoning, the traps to avoid, and the methods that transform raw numbers into an effective argument.

If that seems intimidating, I encourage you to take these skills, embrace them, practice them deliberately, and see how they will serve you in connecting deeply with your audience and producing results that matter. Take small steps, starting with one chart, but start today. The clarity you add to complex information is a valuable contribution to your audience, your organization, and ultimately to the quality of the decisions being made.

Thanks for exploring vertical logic with me! I wish you the best as you continue to hone your visual storytelling abilities, and I am excited to see the impact that you will have through your use of data in the future.

# References

- Begg, I., Armour, V., & Kerr, T. (1985). On believing what we remember. Canadian Journal of Behavioural Science / Revue Canadienne Des Sciences Du Comportement, 17(3), 199–214. doi.org/10.1037/h0080140
- Benoît Mandelbrot. (2006). The fractal geometry of nature. W.H. Freeman and Company.
- Boudry, M., & Braeckman, J. (2012). How convenient! The epistemic rationale of self-validating belief systems. Philosophical Psychology, 25(3), 341–364. doi.org/10.1080/09515089.2011.579420
- Boyd, R. L., Blackburn, K. G., & Pennebaker, J. W. (2020). The narrative arc: Revealing core narrative structures through text analysis. Science Advances, 6(32), eaba2196. doi.org/10.1126/sciadv.aba2196
- Clinton Eye Associates. (n.d.). Color blindness. Clinton Eye Associates. https://www.clintoneye.com/color-blindness.html
- Cothran, H. M., & Wysocki, A. F. (2019). Developing SMART Goals for Your Organization. EDIS, 2005(14). https://doi.org/10.32473/edis-fe577-2005
- Cowan, N. (2010). The Magical Mystery Four: How Is Working Memory Capacity Limited, and Why? Current Directions in Psychological Science, 19(1), 51–57. doi.org/10.1177/0963721409359277
- Fincher, D. (Director). (2008). The Curious Case of Benjamin Button. Paramount Pictures.
- Goldman, A. (1964). Lindy's Law. The New Republic. https://www.gwern.net/docs/statistics/probability/1964-goldman.pdf
- Grigorieva, X. (2015). Pareto optimality in static competitive model of decision-making. Applied Mathematical Sciences, 9, 6217–6223. doi.org/10.12988/ams.2015.56463
- Harari, Y. N., Purcell, J., & Watzman, H. (2018). Sapiens : a brief history of humankind. Harper Perennial.
- Koffka, K. (1935). Principles of Gestalt Psychology. Harcourt, Brace and Company.
- Köhler, W. (1947). Gestalt Psychology: An Introduction to New Concepts in Modern Psychology. Liveright Publishing. (Original work often cited from earlier dates/editions, 1947 is a common accessible English edition).
- Lidwell, W., Holden K., & Butler, J. (2010). Universal Principles of Design: 125 Ways to Enhance Usability, Influence Perception, Increase Appeal, Make Better Design Decisions, and Teach through Design. Rockport Publishers.

- Lerner, J. S., Li, Y., Valdesolo, P., & Kassam, K. S. (2015). Emotion and Decision Making. Annual Review of Psychology, 66(1), 799–823. doi.org/10.1146/annurev-psych-010213-115043
- National Eye Institute. (2019, July 3). Color blindness. Www.nei.nih.gov. https://www.nei.nih.gov/learn-about-eye-health/eye-conditions-and-diseases/color-blindness
- Patterson, K., Grenny, J., Switzler, A., & Mcmillan, R. (2012). Crucial conversations: tools for talking when the stakes are high. Mcgraw-Hill.
- Peterson, L. (2017, November 14). The science behind the art of storytelling. Harvard Business Publishing. https://www.harvardbusiness.org/the-science-behind-the-art-of-storytelling
- Tardi, C. (2022, July 7). The 80-20 rule (aka Pareto Principle): what it Is, how it Works. Investopedia. https://www.investopedia.com/terms/1/80-20-rule.asp#:~:text=What
- Schultz, D. P., & Schultz, S. E. (2015). A History of Modern Psychology (11th ed.). Cengage Learning. (Or a similar reputable History of Psychology textbook).
- Stafford, T., & Grimes, A. (2012). Memory Enhances the Mere Exposure Effect. Psychology & Marketing, 29(12), 995–1003. https://doi.org/10.1002/mar.20581
- Asión-Suñer, L., & López-Forniés, I. (2021). Analysis of Modular Design Applicable in Prosumer Scope. Guideline in the Creation of a New Modular Design Model. Applied Sciences. 11(22), 10620. doi.org/10.3390/app112210620Nassim Nicholas Taleb. 1 (2012). Antifragile: Things That Gain from Disorder. Random House.
- Wertheimer, M. (1912). Experimentelle Studien über das Sehen von Bewegung. Zeitschrift für Psychologie, 61, 161–265
- Winkielman, P., & Cacioppo, J. T. (2001). Mind at ease puts a smile on the face: psychophysiological evidence that processing facilitation elicits positive affect. Journal of Personality and Social Psychology, 81(6), 989–1000. https://pubmed.ncbi.nlm.nih.gov/11761320/

www.ingramcontent.com/pod-product-compliance
Lightning Source LLC
Chambersburg PA
CBHW070932210326
41520CB00021B/6912